PRIVACY AND IDENTITY PROTECTION

# MOBILE DEVICE SECURITY

# THREATS AND CONTROLS

# PRIVACY AND IDENTITY PROTECTION

Additional books in this series can be found on Nova's website under the Series tab.

Additional E-books in this series can be found on Nova's website under the E-book tab.

# MEDIA AND COMMUNICATIONS - TECHNOLOGIES, POLICIES AND CHALLENGES

Additional books in this series can be found on Nova's website under the Series tab.

Additional E-books in this series can be found on Nova's website under the E-book tab.

PRIVACY AND IDENTITY PROTECTION

# MOBILE DEVICE SECURITY THREATS AND CONTROLS

## WILLLIAM R. O'CONNOR
### EDITOR

New York

Copyright © 2013 by Nova Science Publishers, Inc.

**All rights reserved.** No part of this book may be reproduced, stored in a retrieval system or transmitted in any form or by any means: electronic, electrostatic, magnetic, tape, mechanical photocopying, recording or otherwise without the written permission of the Publisher.

For permission to use material from this book please contact us:
Telephone 631-231-7269; Fax 631-231-8175
Web Site: http://www.novapublishers.com

## NOTICE TO THE READER

The Publisher has taken reasonable care in the preparation of this book, but makes no expressed or implied warranty of any kind and assumes no responsibility for any errors or omissions. No liability is assumed for incidental or consequential damages in connection with or arising out of information contained in this book. The Publisher shall not be liable for any special, consequential, or exemplary damages resulting, in whole or in part, from the readers' use of, or reliance upon, this material. Any parts of this book based on government reports are so indicated and copyright is claimed for those parts to the extent applicable to compilations of such works.

Independent verification should be sought for any data, advice or recommendations contained in this book. In addition, no responsibility is assumed by the publisher for any injury and/or damage to persons or property arising from any methods, products, instructions, ideas or otherwise contained in this publication.

This publication is designed to provide accurate and authoritative information with regard to the subject matter covered herein. It is sold with the clear understanding that the Publisher is not engaged in rendering legal or any other professional services. If legal or any other expert assistance is required, the services of a competent person should be sought. FROM A DECLARATION OF PARTICIPANTS JOINTLY ADOPTED BY A COMMITTEE OF THE AMERICAN BAR ASSOCIATION AND A COMMITTEE OF PUBLISHERS.

Additional color graphics may be available in the e-book version of this book.

**Library of Congress Cataloging-in-Publication Data**

ISBN: 978-1-62417-254-0

*Published by Nova Science Publishers, Inc. ✝ New York*

# CONTENTS

| | | |
|---|---|---|
| **Preface** | | **vii** |
| **Chapter 1** | Information Security: Better Implementation of Controls for Mobile Devices Should Be Encouraged<br>*United States Government Accountability Office* | **1** |
| **Chapter 2** | Technical Information Paper: Cyber Threats to Mobile Devices<br>*United States Computer Emergency Readiness Team* | **47** |
| **Chapter 3** | Stolen and Lost Wireless Devices<br>*Federal Communications Commission* | **63** |
| **Chapter 4** | Statement of Jason Weinstein, Deputy Assistant Attorney General, Criminal Division, U.S. Department of Justice. Hearing on "Protecting Mobile Privacy: Your Smartphones, Tablets, Cell Phones and Your Privacy" | **67** |
| **Chapter 5** | Statement of Justin Brookman, Director, Consumer Privacy, Center for Democracy and Technology. Hearing on "Protecting Mobile Privacy: Your Smartphones, Tablets, Cell Phones and Your Privacy" | **79** |
| **Index** | | **93** |

# PREFACE

Millions of Americans currently use mobile devices (e.g., cellphones, smartphones, and tablet computers) on a daily basis to communicate, obtain internet-based information, and share their own information, photographs, and videos. Given the extent of consumer reliance on mobile interactions, it is increasingly important that these devices be secured from expanding threats to the confidentiality, integrity, and availability of the information they maintain and share. This book explores the current challenge of threats and controls to mobile device security with a focus on cyber threats to mobile devices; wireless device theft; and how to protect the data on your phone.

Chapter 1 – Millions of Americans currently use mobile devices—e.g., cellphones, smartphones, and tablet computers— on a daily basis to communicate, obtain Internet-based information, and share their own information, photographs, and videos. Given the extent of consumer reliance on mobile interactions, it is increasingly important that these devices be secured from expanding threats to the confidentiality, integrity, and availability of the information they maintain and share.

Accordingly, GAO was asked to determine (1) what common security threats and vulnerabilities affect mobile devices, (2) what security features and practices have been identified to mitigate the risks associated with these vulnerabilities, and (3) the extent to which government and private entities have been addressing the security vulnerabilities of mobile devices. To do so, GAO analyzed publically available mobile security reports, surveys related to consumer cybersecurity practices, as well as statutes, regulations, and agency policies; GAO also interviewed representatives from federal agencies and private companies with responsibilities in telecommunications and cybersecurity.

Chapter 2 – Today's advanced mobile devices are well integrated with the Internet and have far more functionality than mobile phones of the past. They are increasingly used in the same way as personal computers (PCs), potentially making them susceptible to similar threats affecting PCs connected to the Internet. Since mobile devices can contain vast amounts of sensitive and personal information, they are attractive targets that provide unique opportunities for criminals intent on exploiting them. Both individuals and society as a whole can suffer serious consequences if these devices are compromised.

This paper introduces emerging threats likely to have a significant impact on mobile devices and their users.

Chapter 3 – The theft of wireless devices, particularly smartphones, is sharply on the rise across the country, according to many published reports. The high resale value of these high-tech phones has made them a prime target for robbers and the personal information contained on the device that could be used by identity thieves. Below are several steps that you can take to better protect yourself, your device, and the data it contains, along with instructions on what to do if your device is lost or stolen.

Chapter 4 – This is Statement of Jason Weinstein.

Chapter 5 – This is Statement of Justin Brookman.

In: Mobile Device Security
Editor: Willliam R. O'Connor

ISBN: 978-1-62417-254-0
© 2013 Nova Science Publishers, Inc.

*Chapter 1*

# INFORMATION SECURITY: BETTER IMPLEMENTATION OF CONTROLS FOR MOBILE DEVICES SHOULD BE ENCOURAGED[*]

## United States Government Accountability Office

### WHY GAO DID THIS STUDY

Millions of Americans currently use mobile devices—e.g., cellphones, smartphones, and tablet computers— on a daily basis to communicate, obtain Internet-based information, and share their own information, photographs, and videos. Given the extent of consumer reliance on mobile interactions, it is increasingly important that these devices be secured from expanding threats to the confidentiality, integrity, and availability of the information they maintain and share.

Accordingly, GAO was asked to determine (1) what common security threats and vulnerabilities affect mobile devices, (2) what security features and practices have been identified to mitigate the risks associated with these vulnerabilities, and (3) the extent to which government and private entities have been addressing the security vulnerabilities of mobile devices. To do so,

---

[*] This is an edited, reformatted and augmented version of United States Government Accountability Office, Publication No. GAO-12-757, dated September 2012.

GAO analyzed publically available mobile security reports, surveys related to consumer cybersecurity practices, as well as statutes, regulations, and agency policies; GAO also interviewed representatives from federal agencies and private companies with responsibilities in telecommunications and cybersecurity.

## WHAT GAO RECOMMENDS

GAO recommends that FCC encourage the private sector to implement a broad, industry-defined baseline of mobile security safeguards. GAO also recommends that DHS and NIST take steps to better measure progress in raising national cybersecurity awareness. The FCC, DHS, and NIST generally concurred with GAO's recommendations.

## WHAT GAO FOUND

Threats to the security of mobile devices and the information they store and process have been increasing significantly. For example, the number of variants of malicious software, known as "malware," aimed at mobile devices has reportedly risen from about 14,000 to 40,000 or about 185 percent in less than a year. Cyber criminals may use a variety of attack methods, including intercepting data as they are transmitted to and from mobile devices and inserting malicious code into software applications to gain access to users' sensitive information. These threats and attacks are facilitated by vulnerabilities in the design and configuration of mobile devices, as well as the ways consumers use them. Common vulnerabilities include a failure to enable password protection and operating systems that are not kept up to date with the latest security patches.

Mobile device manufacturers and wireless carriers can implement technical features, such as enabling passwords and encryption to limit or prevent attacks. In addition, consumers can adopt key practices, such as setting passwords and limiting the use of public wireless connections for sensitive transactions, which can significantly mitigate the risk that their devices will be compromised.

Federal agencies and private companies have promoted secure technologies and practices through standards and public private partnerships.

Despite these efforts, safeguards have not been consistently implemented. Although the Federal Communications Commission (FCC) has facilitated public-private coordination to address specific challenges such as cellphone theft, it has not yet taken similar steps to encourage device manufacturers and wireless carriers to implement a more complete industry baseline of mobile security safeguards. In addition, many consumers still do not know how to protect themselves from mobile security vulnerabilities, raising questions about the effectiveness of public awareness efforts. The Department of Homeland Security (DHS) and National Institute of Standards and Technology (NIST) have not yet developed performance measures or a baseline understanding of the current state of national cybersecurity awareness that would help them determine whether public awareness efforts are achieving stated goals and objectives.

## ABBREVIATIONS

| | |
|---|---|
| Commerce | Department of Commerce |
| CSRIC | Communications, Security, Reliability, and Interoperability Council |
| DHS | Department of Homeland Security |
| DOD | Department of Defense |
| FCC | Federal Communications Commission |
| FISMA | Federal Information Security Management Act |
| FTC | Federal Trade Commission |
| http | hypertext transfer protocol |
| NCSA | National Cyber Security Alliance |
| NICE | National Initiative for Cybersecurity Education |
| NIST | National Institute of Standards and Technology |
| NTIA | National Telecommunications and Information Administration |
| OMB | Office of Management and Budget |
| PIN | personal identification number |
| PKI | public key infrastructure |
| US-CERT | US-Computer Emergency Readiness Team |
| VPN | virtual private network |

The Honorable Fred Upton Chairman

The Honorable Henry Waxman
Ranking Member
Committee on Energy and Commerce
House of Representatives

The Honorable Greg Walden
Chairman

The Honorable Anna Eshoo
Ranking Member
Committee on Energy and Commerce
Subcommittee on Communications and Technology
House of Representatives

The Honorable Cliff Stearns
Chairman

The Honorable Diana DeGette
Ranking Member
Committee on Energy and Commerce
Subcommittee on Oversight and Investigations
House of Representatives

Millions of Americans currently use mobile devices—cellphones, smartphones, and tablet computers—on a daily basis to communicate, obtain Internet-based information, and share information, photographs, and videos. Dramatic recent advances in the technical capabilities of mobile devices have paved the way for increased connectivity. As a result, consumers can now carry out a broad range of interactions, including sensitive transactions, which previously required the use of a desktop or laptop computer. Given the extent of consumer reliance on mobile interactions, it is increasingly important that these devices be secured from threats to the confidentiality, integrity, and availability of the information they maintain and share.

Accordingly, our objectives were to determine: (1) what common security threats and vulnerabilities affect mobile devices, (2) what security features and practices have been identified to mitigate the risks associated with these vulnerabilities, and (3) the extent to which government and private entities have been addressing the security vulnerabilities of mobile devices. To assess common security threats and vulnerabilities as well as security controls and practices to address them, we obtained and reviewed published analyses and guidance, including databases of mobile security vulnerabilities.

We also interviewed representatives from federal agencies and private companies with responsibilities in the telecommunications and cybersecurity fields to obtain current information about threats and vulnerabilities.

To determine the extent to which the government and private entities are addressing vulnerabilities, we analyzed statutes, regulations, agency policies, and technical standards.

We also interviewed officials from federal agencies and private companies to identify actions taken to address mobile security, such as developing guidance and sharing information.

We conducted this performance audit from November 2011 to September 2012 in accordance with generally accepted government auditing standards. Those standards require that we plan and perform the audit to obtain sufficient, appropriate evidence to provide a reasonable basis for our findings and conclusions based on our audit objectives.

We believe that the evidence obtained provides a reasonable basis for our findings and conclusions based on our audit objectives. Appendix I contains additional details on the objectives, scope, and methodology of our review.

## BACKGROUND

Consumer adoption of mobile devices is growing rapidly, enabled by affordable prices, increasingly reliable connections, and faster transmission speeds. According to a recent analysis, mobile devices are the fastest growing consumer technology, with worldwide sales increasing from 300 million in 2010 to an estimated 650 million in 2012.[1]

Advances in computing technology have resulted in increased speed and storage capacity for mobile devices. The advances have enhanced consumers' abilities to perform a wide range of online tasks.

While these devices provide many productivity benefits to consumers and organizations, they also pose security risks if not properly protected.

## A Variety of Entities Provide Products and Services to Consumers

Several different types of private sector entities provide products and services that are used by consumers as part of a seamless mobile telecommunications system. These entities include mobile device manufacturers, operating system developers, application developers, and wireless carriers.

### *Mobile Device Manufacturers*

Manufacturers of mobile equipment include both hardware and software developers. Components of the hardware or software on any given device may come from multiple manufacturers. Major device manufacturers with the largest total market shares in the United States include Apple Inc., HTC Corporation, Research In Motion, Corp., Motorola Mobility Inc., Samsung, and LG Electronics. The products they develop include cellphones, smartphones, and tablet computers.

- A cellphone is a device that can make and receive telephone calls over a radio network while moving around a wide geographic area. According to a recent report,[2] 88 percent of American adults owned cellphones as of February 2012.
- A smartphone has more capabilities than a cellphone. Consumers can use smartphones to run a wide variety of general and special-purpose software applications. Smartphones typically have a larger graphical display with greater resolution than cellphones and have either a keyboard or touch-sensitive screen for alphanumeric input. Smartphones also offer expansion capabilities and other built-in wireless communications (such as WiFi and Bluetooth services). [3] According to a recent report,[4] 46 percent of American adults owned smartphones as of February 2012.
- A tablet personal computer is a portable personal computer with a touch-sensitive screen. The tablet form is typically smaller than a notebook computer but larger than a smartphone. According to a recent report,[5] 19 percent of American adults owned a tablet as of January 2012.

## Mobile Operating System Developers

Operating system developers build the software that provides basic computing functions and controls for mobile devices. The operating system is the software platform used by other programs, called applications, to interact with the mobile device. Major operating system developers for mobile devices with the largest total market shares in the United States include Apple Inc., Google Inc., and Research In Motion, Corp.

- **Apple Inc.** The mobile operating system developed and distributed by Apple is known as iOS. It is a proprietary system; all updates and other changes to the software are administered by Apple. In addition, all software applications that run on iOS devices (e.g., iPhones and iPads) are required to conform to specifications established by Apple and be digitally signed by approved developers. Apple distributes these applications through its online "store," called App Store.
- **Google Inc.** As a member of the Open Handset Alliance,[6] Google Inc. led the development of Android, an operating system for mobile devices, based on the Linux operating system. Android, like Linux, is an "open" operating system, meaning that its software code is publicly available and can be tailored to the needs of individual devices and telecommunications carriers. Thus, many different tailored versions of the software are in use. To run on Android devices, software applications need to be digitally signed by the developer, who is responsible for the application's behavior. Android applications are made available on third-party application marketplaces, websites, and on the online official Android application store called Google Play.
- **Research In Motion, Corp.** Research In Motion developed a proprietary operating system for its BlackBerry mobile devices. Although a proprietary system, it can run any third-party applications that are written in Java. [7] Applications are tested by Research In Motion before users can download them. In addition, any application given access to sensitive data or features when installed is required by Research In Motion to be digitally signed by the developer. BlackBerry applications are available for download on the online store called BlackBerry App World.

## Mobile Application Developers

Mobile application developers develop the software applications that consumers interact with directly. In many cases, these applications provide the same services that are available through traditional websites, such as news and information services, online banking, shopping, and electronic games. Other applications are designed to take into account a user's physical location to provide tailored information or services, such as information about nearby shops, restaurants, or other elements of the physical environment.

*Wireless Carriers*

Wireless carriers manage telecommunications networks and provide phone services, including mobile devices, directly to consumers. While carriers do not design or manufacture their own mobile devices, in some cases they can influence the design and the features of other manufacturers' products because they control sales and interactions with large numbers of consumers. Major wireless carriers with the largest total market shares in the United States include Verizon Wireless, AT&T Inc., Sprint, and T-Mobile USA Inc.

Carriers provide basic telephone service through wireless cellular networks which cover large distances. However, other types of shorter-range wireless networks may also be used with mobile devices. These shorter-range networks may be supported by the same carriers or by different providers. Major types of wireless networks include cellular networks, WiFi networks, and wireless personal area networks.

- **Cellular networks.** Cellular networks are managed by carriers and provide coverage based on dividing a large geographical service area into smaller areas of coverage called "cells." The cellular network is a radio network distributed over the cells and each cell has a base station equipped with an antenna to receive and transmit radio signals to mobile phones within its coverage area. A mobile device's communications are generally associated with the base station of the cell in which it is located. Each base station is linked to a mobile telephone switching office, which is also connected to the local wireline telephone network. The mobile phone switching office directs calls to the desired locations, whether to another mobile phone or a traditional wireline telephone. This office is responsible for switching calls from one cell to another in a smooth and seamless manner as consumers change locations during a call. Figure 1 depicts the key components of this cellular network.

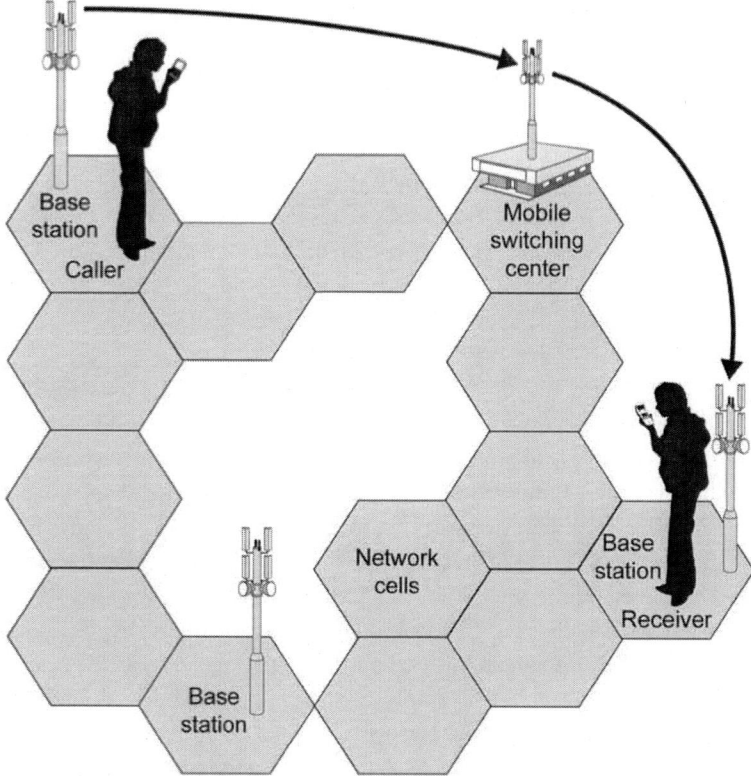

Source: GAO.

Figure 1. Key Components of a Cellular Network.

- **WiFi networks.** WiFi networking nodes may be established by businesses or consumers to provide networking service within a limited geographic area, such as within a home, office, or place of business. They are generally composed of two basic elements: access points and wireless-enabled devices, such as smart phones and tablet computers. These devices use radio transmitters and receivers to communicate with each other. Access points are physically wired to a conventional network and provide a means for wireless devices to connect to them. WiFi networks conform to the Institute of Electrical and Electronics Engineers 802.11 standards.[8]
- **Other wireless personal area networks.** Other wireless personal area networks may be used that do not conform to the WiFi standard.

For example, the Bluetooth standard[9] is often used to establish connectivity with nearby components, such as headsets or computer keyboards.

## Federal Agencies Have Roles in Addressing Mobile Security

While federal agencies are not responsible for ensuring the security of individual mobile devices, several are involved in activities designed to address and promote cybersecurity and mobile security in general.

- The Department of Commerce (Commerce) is responsible under Homeland Security Presidential Directive 7[10] in coordination with other federal and nonfederal entities, for improving technology for cyber systems and promoting efforts to protect critical infrastructure. Within Commerce, the National Institute of Standards and Technology (NIST) is responsible for developing information security standards and guidelines, including minimum requirements for unclassified federal information systems, as part of its statutory responsibilities under the Federal Information Security Management Act (FISMA).[11] For example, NIST has developed guidelines on cellphone and Bluetooth security.[12] These standards and guidelines are generally made available to the public and can be used by both the public and private sectors. NIST also serves as the lead federal agency for coordinating the National Initiative for Cybersecurity Education (NICE) with other agencies. According to NIST, NICE seeks to establish an operational, sustainable, and continually improving cybersecurity education program for the nation. NICE includes an awareness initiative, which is led by the Department of Homeland Security (DHS), which focuses on boosting national cybersecurity awareness through public service campaigns to promote cybersecurity and responsible use of the Internet, and making cybersecurity a popular educational and career pursuit for older students. As we previously reported,[13] NIST developed a draft strategic plan for the NICE initiative. This plan includes strategic goals, supporting objectives, and related activities for the awareness component. Specifically, the draft strategic plan calls for (1) improving citizens' knowledge to allow them to make smart choices as they manage online risk, (2) improving knowledge of cybersecurity within organizations so that resources are well applied to meet the most obvious and serious threats, and (3) enabling access to cybersecurity resources. The plan also identifies supporting activities and products designed to support the overarching goal, such as the "Stop. Think. Connect." awareness campaign.

According to Commerce's National Telecommunications and Information Administration (NTIA), it serves as the President's principal adviser on telecommunications policies pertaining to economic and technological advancement and to the regulation of the telecommunications industry, including mobile telecommunications. NTIA is responsible for coordinating telecommunications activities of the executive branch and assisting in the formulation of policies and standards for those activities, including considerations of interoperability, privacy, security, spectrum use, and emergency readiness.

- Federal law and policy tasks DHS with critical infrastructure protection responsibilities that include creating a safe, secure, and resilient cyber environment in conjunction with other federal agencies, other levels of government, international organizations, and industry. The National Strategy to Secure Cyberspace tasked DHS as the lead agency in promoting a comprehensive national awareness program to empower Americans to secure their own parts of cyberspace.[14] Consistent with that tasking, DHS is currently leading the awareness component of NICE.

- The Federal Communications Commission's (FCC) role in mobile security stems from its broad authority to regulate interstate and international communications, including for the purpose of "promoting safety of life and property."[15] In addition, FCC has established the Communications, Security, Reliability, and Interoperability Council (CSRIC). CSRIC is a federal advisory committee whose mission is to provide recommendations to FCC to help ensure, among other things, secure and reliable communications systems, including telecommunications, media, and public safety. A previous CSRIC[16] included a working group that was focused on identifying cybersecurity best practices (including mobile security practices), and had representation from segments of the communications industry and public safety communities. The current CSRIC has focused on the development and implementation of best practices related to several specific cybersecurity topics. FCC has also

established a Technological Advisory Council, which includes various working groups, one of which has been working since March 2012 to identify, prioritize, and analyze mobile security and privacy issues.
- The Federal Trade Commission (FTC) promotes competition and protects the public by, among other things, bringing enforcement actions against entities that engage in unfair or deceptive acts or practices.[17] An unfair act is an act or practice that causes or is likely to cause substantial injury to consumers that is not reasonably avoidable by consumers and is not outweighed by countervailing benefits to consumers or to competition. A deceptive act or practice occurs if there is a representation, omission, or practice that is likely to mislead the consumer acting reasonably in the circumstances, to the consumer's detriment. According to FTC, its authority to bring enforcement actions covers many of the entities that provide mobile products and services to consumers, including mobile device manufacturers, operating system developers, and application developers. FTC's jurisdiction also extends to wireless carriers when they are not engaged in common carrier activities. For example, mobile phone operators engaging in mobile payments functions such as direct-to-carrier billing are under FTC's jurisdiction.
- The Department of Defense (DOD) is responsible for security systems, including mobile devices that use its networks or contain DOD data. While it has no responsibility with regards to consumer mobile devices, its guidance can be useful for consumers. For example, the DOD Security Technical Implementation Guides are available to the public. These guides contain technical guidance to secure information systems or software that might otherwise be vulnerable to a malicious computer attack. In addition, certain guides address aspects of mobile device security.
- The Office of Management and Budget (OMB) is responsible for overseeing and providing guidance to federal agencies on the use of information technology, which can include mobile devices. One OMB memorandum to federal agencies, for example, instructs agencies to properly safeguard information stored on federal systems (including mobile devices) by requiring the use of encryption and a "time-out" function for re-authentication after 30 minutes of inactivity.[18]

# MOBILE DEVICES FACE A BROAD RANGE OF SECURITY THREATS AND VULNERABILITIES

Threats[19] to the security of mobile devices and the information they store and process have been increasing significantly.[20] Many of these threats are similar to those that have long plagued traditional computing devices connected to the Internet.

For example, cyber criminals and hackers have a variety of attack methods readily available to them, including using software tools to intercept data as they are transmitted to and from a mobile device, inserting malicious software code into the operating systems of mobile devices by including it in seemingly harmless software applications, and using e-mail phishing techniques to gain access to mobile-device users' sensitive information.

The significance of these threats, which are growing in number and kind, is magnified by the vulnerabilities associated with mobile devices. Common vulnerabilities[21] in mobile devices include a failure to enable password protection, the lack of the capability to intercept malware, and operating systems that are not kept up to date with the latest security patches.

## Attacks on Mobile Devices Are Increasing

Cyber-based attacks against mobile devices are evolving and increasing. Examples of recent incidents include:

- In May 2012, a regulatory agency in the United Kingdom fined a company for distributing malware versions of popular gaming applications that triggered mobile devices to send costly text messages to a premium-rate telephone number.
- In February 2012, a cybersecurity firm, Symantec Corporation, reported that a large number of Android devices in China were infected with malware that connected them to a botnet.[22] The botnet's operator was able to remotely control the devices and incur charges on user accounts for premium services such as sending text messages to premium numbers, contacting premium telephony services, and connecting to pay-per-view video services. The number of infected devices able to generate revenue on any given day ranged from

10,000 to 30,000, enough to potentially net the botnet's operator millions of dollars annually if infection rates were sustained.
- In January 2012, an antivirus company reported that hackers had subverted the search results for certain popular mobile applications so that they would redirect users to a web page where they were encouraged to download a fake antivirus program containing malware.
- In October 2011, FTC reached a settlement of an unfair practice case with a company after alleging that its mobile application was likely to cause consumers to unwittingly disclose personal files, such as pictures and videos, stored on their smartphones and tablet computers. The company had configured the application's default settings so that upon installation and set-up it would publicly share users' photos, videos, documents, and other files stored on those devices.

These incidents reflect a trend of increasing global attacks against mobile devices. Specifically, recent studies have found that

- mobile malware grew by 155 percent in 2011;[23]
- new mobile vulnerabilities have been increasing, from 163 in 2010 to 315 in 2011, an increase of over 93 percent;[24]
- an estimated half million to one million people had malware on their Android devices in the first half of 2011; and[25]
- 3 out of 10 Android owners are likely to encounter a threat on their device each year as of 2011.[26]

According to a networking technology company, Juniper Networks, malware aimed at mobile devices is increasing. For example, the number of variants of malicious software, known as "malware," aimed at mobile devices has reportedly risen from about 14,000 to 40,000, a 185 percent increase in less than a year. Figure 2 shows the increase in malware variants between July 2011 and May 2012.

The increasing prevalence of attacks against mobile devices makes it important to assess and understand the nature of the threats they face and the vulnerabilities these attacks exploit.

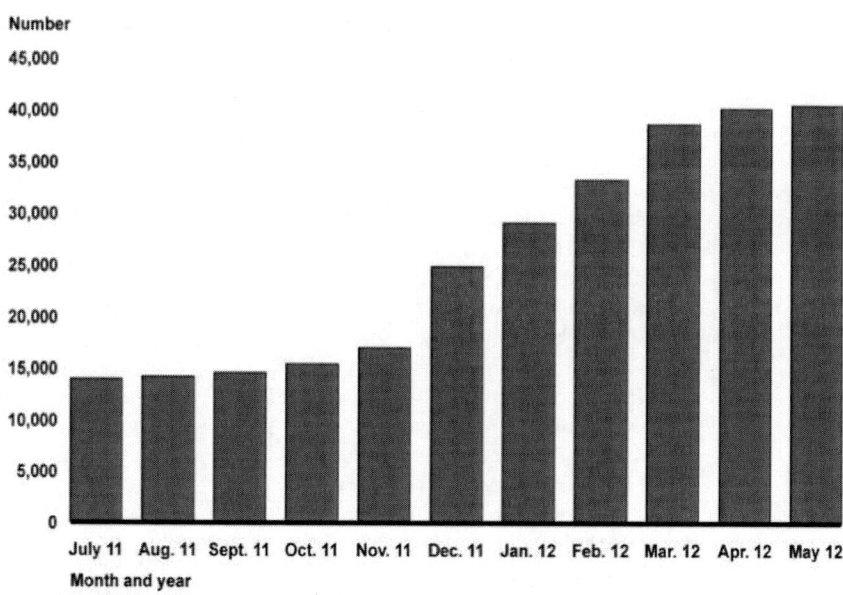

Source: GAO based on Juniper Networks, Inc.

Figure 2. Number of Malware Variants Identified Globally between July 2011 and May 2012.

## Sources of Threats and Attack Methods Vary

Mobile devices face a range of cybersecurity threats. These threats can be unintentional or intentional. Unintentional threats can be caused by software upgrades or defective equipment that inadvertently disrupt systems. Intentional threats include both targeted and untargeted attacks from a variety of sources, including botnet operators, cyber criminals, hackers, foreign nations engaged in espionage, and terrorists.

These threat sources vary in terms of the capabilities of the actors, their willingness to act, and their motives, which can include monetary gain or political advantage, among others. For example, cyber criminals are using various attack methods to access sensitive information stored and transmitted by mobile devices.

Table 1 summarizes those groups or individuals that are key sources of threats for mobile devices.

**Table 1. Sources of Mobile Threats**

| Threat source | Description |
|---|---|
| Botnet operators | Botnet operators use malware distributed to large numbers of mobile devices and other electronic systems to coordinate remotely controlled attacks on websites and to distribute phishing schemes, spam, and further malware attacks on individual mobile devices. |
| Cyber criminals | Cyber criminals generally attack mobile devices for monetary gain. They may use spam, phishing, and spyware/malware to gain access to the information stored on a device, which they then use to commit identity theft, online fraud, and computer extortion.<br><br>In addition, international criminal organizations pose a threat to corporations, government agencies, and other institutions by attacking mobile devices to conduct industrial espionage and large-scale monetary and intellectual property theft. |
| Foreign governments | Foreign intelligence services may attack mobile devices as part of their information-gathering and espionage activities.<br><br>Foreign governments may develop information warfare doctrine, programs, and capabilities that could disrupt the supply chain, mobile communications, and economic infrastructures thatsupport homeland security and national defense. |
| Hackers | Hackers may attack mobile devices to demonstrate their skill or gain prestige in the hacker community.<br><br>While hacking once required a fair amount of skill or computer knowledge, hackers can now download attack scripts and protocols from the Internet and easily launch them against mobile devices. |
| Terrorists | Terrorists may seek to destroy, incapacitate, or exploit critical infrastructures such as mobile networks, to threaten national security, weaken the U.S. economy, or damage public morale and confidence. Terrorists may also use phishing schemes or spyware/malware to generate funds or gather sensitive information from mobile devices. |

Source: GAO analysis based on data from the Director of National Intelligence, Department of Justice, Central Intelligence Agency, the Software Engineering Institute's CERT® Coordination Center, and security reports.

These threat sources may use a variety of techniques, or exploits, to gain control of mobile devices or to access sensitive information on them. Common mobile attacks are presented in table 2.

## Table 2. Common Mobile Attacks

| Attacks | Description |
|---|---|
| Browser exploits | These exploits are designed to take advantage of vulnerabilities in software used to access websites. Visiting certain web pages and/or clicking on certain hyperlinks can trigger browser exploits that install malware or perform other adverse actions on a mobile device. |
| Data interception | Data interception can occur when an attacker is eavesdropping on communications originating from or being sent to a mobile device. Electronic eavesdropping is possible through various techniques, such as (1) man-in-the-middle attacks, which occur when a mobile device connects to an unsecured WiFi network and an attacker intercepts and alters the communication; and (2) WiFi sniffing, which occurs when data are sent to or from a device over an unsecured (i.e., not encrypted) network connection, allowing an eavesdropper to "listen to" and record the information that is exchanged. |
| Keystroke logging | This is a type of malware that records keystrokes on mobile devices in order to capture sensitive information, such as credit card numbers. Generally keystroke loggers transmit the information they capture to a cyber criminal's website or e-mail address. |
| Malware | Malware is often disguised as a game, patch, utility, or other useful third-party software application. Malware can include spyware (software that is secretly installed to gather information on individuals or organizations without their knowledge), viruses (a program that can copy itself and infect the mobile system without permission or knowledge of the user), and Trojans (a type of malware that disguises itself as or hides itself within a legitimate file). Once installed, malware can initiate a wide range of attacks and spread itself onto other devices. The malicious application can perform a variety of functions, including accessing location information and other sensitive information, gaining read/write access to the user's browsing history, as well as initiating telephone calls, activating the device's microphone or camera to surreptitiously record information, and downloading other malicious applications. Repackaging—the process of modifying a legitimate application to insert malicious code—is one technique that an attacker can use. |
| Unauthorized location tracking | Location tracking allows the whereabouts of registered mobile devices to be known and monitored. While it can be done openly for legitimate purposes, it may also take place surreptitiously. Location data may be obtained through legitimate software applications as well as malware loaded on the user's mobile device. |

**Table 2. (Continued)**

| Attacks | Description |
|---|---|
| Network exploits | Network exploits take advantage of software flaws in the system that operates on local (e.g., Bluetooth, WiFi) or cellular networks. Network exploits often can succeed without any user interaction, making them especially dangerous when used to automatically propagate malware. With special tools, attackers can find users on a WiFi network, hijack the users' credentials, and use those credentials to impersonate a user online. Another possible attack, known as bluesnarfing, enables attackers to gain access to contact data by exploiting a software flaw in a Bluetooth-enabled device. |
| Phishing | Phishing is a scam that frequently uses e-mail or pop-up messages to deceive people into disclosing sensitive information. Internet scammers use e-mail bait to "phish" for passwords and financial information from mobile users and other Internet users. |
| Spamming | Spam is unsolicited commercial e-mail advertising for products, services, and websites. Spam canalso be used as a delivery mechanism for malicious software. Spam can appear in text messages as well as electronic mail. Besides the inconvenience of deleting spam, users may face charges for unwanted text messages. Spam can also be used for phishing attempts. |
| Spoofing | Attackers may create fraudulent websites to mimic or "spoof" legitimate sites and in some cases may use the fraudulent sites to distribute malware to mobile devices. E-mail spoofing occurs whenthe sender address and other parts of an e-mail header are altered to appear as though the e-mailoriginated from a different source. Spoofing hides the origin of an e-mail message. Spoofed e-mails may contain malware. |
| Theft/loss | Because of their small size and use outside the office, mobile devices can be easier to misplace or steal than a laptop or notebook computer. If mobile devices are lost or stolen, it may be relatively easy to gain access to the information they store. |
| Zero-day exploit | A zero-day exploit takes advantage of a security vulnerability before an update for the vulnerabilityis available. By writing an exploit for an unknown vulnerability, the attacker creates a potential threat because mobile devices generally will not have software patches to prevent the exploit fromsucceeding. |

Source: GAO analysis of data from the National Institute of Standards and Technology, United States Computer Emergency Readiness Team, and industry reports.

Attacks against mobile devices generally occur through four different channels of activities:

- **Software downloads.** Malicious applications may be disguised as a game, device patch, or utility, which is available for download by unsuspecting users and provides the means for unauthorized users to gain unauthorized use of mobile devices and access to private information or system resources on mobile devices.
- **Visiting a malicious website.** Malicious websites may automatically download malware to a mobile device when a user visits. In some cases, the user must take action (such as clicking on a hyperlink) to download the application, while in other cases the application may download automatically.
- **Direct attack through the communication network.** Rather than targeting the mobile device itself, some attacks try to intercept communications to and from the device in order to gain unauthorized use of mobile devices and access to sensitive information.
- **Physical attacks.** Unauthorized individuals may gain possession of lost or stolen devices and have unauthorized use of mobile devices and access sensitive information stored on the device.

## A Range of Vulnerabilities Facilitate Attacks

Mobile devices are subject to numerous security vulnerabilities, including a failure to enable password protection, the inability to intercept malware, and operating systems that are not kept up to date with the latest security patches. While not a comprehensive list of all possible vulnerabilities, the following 10 vulnerabilities can be found on all mobile platforms.

- **Mobile devices often do not have passwords enabled.** Mobile devices often lack passwords to authenticate users and control access to data stored on the devices. Many devices have the technical capability to support passwords, personal identification numbers (PIN), or pattern screen locks for authentication. Some mobile devices also include a biometric reader to scan a fingerprint for authentication. However, anecdotal information indicates that consumers seldom employ these mechanisms. Additionally, if users do use a password or PIN they often choose passwords or PINs that can be easily determined or bypassed, such as 1234 or 0000. Without passwords or PINs to lock the device, there is increased risk that stolen or lost

phones' information could be accessed by unauthorized users who could view sensitive information and misuse mobile devices.
- **Two-factor authentication is not always used when conducting sensitive transactions on mobile devices.** According to studies, consumers generally use static passwords instead of two-factor authentication when conducting online sensitive transactions while using mobile devices. Using static passwords for authentication has security drawbacks: passwords can be guessed, forgotten, written down and stolen, or eavesdropped. Two-factor authentication generally provides a higher level of security than traditional passwords and PINs, and this higher level may be important for sensitive transactions. Two-factor refers to an authentication system in which users are required to authenticate using at least two different "factors"—something you know, something you have, or something you are—before being granted access. Mobile devices themselves can be used as a second factor in some two-factor authentication schemes. The mobile device can generate pass codes, or the codes can be sent via a text message to the phone. Without two-factor authentication, increased risk exists that unauthorized users could gain access to sensitive information and misuse mobile devices.
- **Wireless transmissions are not always encrypted.** Information such as e-mails sent by a mobile device is usually not encrypted while in transit. In addition, many applications do not encrypt the data they transmit and receive over the network, making it easy for the data to be intercepted. For example, if an application is transmitting data over an unencrypted WiFi network using hypertext transfer protocol (http) (rather than secure http),[27] the data can be easily intercepted. When a wireless transmission is not encrypted, data can be easily intercepted by eavesdroppers, who may gain unauthorized access to sensitive information.
- **Mobile devices may contain malware.** Consumers may download applications that contain malware. Consumers download malware unknowingly because it can be disguised as a game, security patch, utility, or other useful application. It is difficult for users to tell the difference between a legitimate application and one containing malware. For example, figure 3 shows how an application could be repackaged with malware and a consumer could inadvertently download it onto a mobile device.

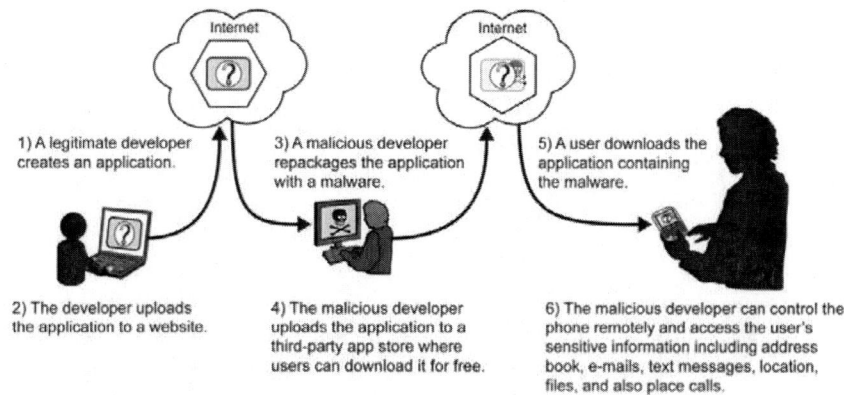

Source: GAO analysis of students and security reports.

Figure 3. Repackaging Applications with Malware.

- **Mobile devices often do not use security software.** Many mobile devices do not come preinstalled with security software to protect against malicious applications, spyware, and malware-based attacks. Further, users do not always install security software, in part because mobile devices often do not come preloaded with such software. While such software may slow operations and affect battery life on some mobile devices, without it, the risk may be increased that an attacker could successfully distribute malware such as viruses, Trojans, spyware, and spam, to lure users into revealing passwords or other confidential information.
- **Operating systems may be out-of-date.** Security patches or fixes for mobile devices' operating systems are not always installed on mobile devices in a timely manner. It can take weeks to months before security updates are provided to consumers' devices. Depending on the nature of the vulnerability, the patching process may be complex and involve many parties. For example, Google develops updates to fix security vulnerabilities in the Android OS, but it is up to device manufacturers to produce a device-specific update incorporating the vulnerability fix, which can take time if there are proprietary modifications to the device's software. Once a manufacturer produces an update, it is up to each carrier to test it and transmit the updates to consumers' devices. However, carriers can be delayed in providing the updates because they need time to test whether they interfere with other aspects of the device or the software installed on it.

In addition, mobile devices that are older than 2 years may not receive security updates because manufacturers may no longer support these devices. Many manufacturers stop supporting smartphones as soon as 12 to 18 months after their release. Such devices may face increased risk if manufacturers do not develop patches for newly discovered vulnerabilities.

- **Software on mobile devices may be out-of-date.** Security patches for third-party applications are not always developed and released in a timely manner. In addition, mobile third-party applications, including web browsers, do not always notify consumers when updates are available. Unlike traditional web browsers, mobile browsers rarely get updates. Using outdated software increases the risk that an attacker may exploit vulnerabilities associated with these devices.
- **Mobile devices often do not limit Internet connections.** Many mobile devices do not have firewalls to limit connections. When the device is connected to a wide area network it uses communications ports to connect with other devices and the Internet. These ports are similar to doorways to the device. A hacker could access the mobile device through a port that is not secured. A firewall secures these ports and allows the user to choose what connections he or she wants to allow into the mobile device. The firewall intercepts both incoming and outgoing connection attempts and blocks or permits them based on a list of rules. Without a firewall, the mobile device may be open to intrusion through an unsecured communications port, and an intruder may be able to obtain sensitive information on the device and misuse it.
- **Mobile devices may have unauthorized modifications.** The process of modifying a mobile device to remove its limitations so consumers can add additional features (known as "jailbreaking" or "rooting") changes how security for the device is managed and could increase security risks. Jailbreaking allows users to gain access to the operating system of a device so as to permit the installation of unauthorized software functions and applications and/or to not be tied to a particular wireless carrier. While some users may jailbreak or root their mobile devices specifically to install security enhancements such as firewalls, others may simply be looking for a less expensive or easier way to install desirable applications. In the latter case, users face increased security risks, because they are bypassing the application vetting process established by the manufacturer and thus

The number and variety of threats aimed at mobile devices combined with the vulnerabilities in the way the devices are configured and used by consumers means that consumers face significant risks that the proper functioning of their devices as well as the sensitive information contained on them could be compromised.

# SECURITY CONTROLS AND PRACTICES IDENTIFIED BY EXPERTS CAN REDUCE VULNERABILITIES

Mobile device manufacturers and wireless carriers can implement a number of technical features, such as enabling passwords and encryption, to limit or prevent attacks. In addition, consumers can adopt key practices, such as setting passwords, installing software to combat malware, and limiting the use of public wireless connections for sensitive transactions, which also can significantly mitigate the risk that their devices will be compromised.

## Security Controls for Mobile Devices

Table 3 outlines security controls that can be enabled on mobile devices to help protect against common security threats and vulnerabilities. The security controls and practices described are not a comprehensive list, but are consistent with recent studies[28] and guidance from NIST and DHS, as well as recommended practices identified by the FCC CSRIC advisory committee. In addition, security experts, device manufacturers, and wireless carriers agreed that the security controls and practices identified are comprehensive and are in agreement with the lists.

Appendix II provides links to federal websites that provide information on mobile security.

**Table 3. Key Security Controls to Combat Common Threats and Vulnerabilities**

| Security control | Description |
|---|---|
| Enable user authentication | Devices can be configured to require passwords or PINs to gain access. In addition, the password field can be masked to prevent it from being observed, and the devices can activate idle-time screen locking to prevent unauthorized access. |

have less protection against inadvertently installing malware. Further, jailbroken devices may not receive notifications of security updates from the manufacturer and may require extra effort from the user to maintain up-to-date software.

- **Communication channels may be poorly secured.** Having communication channels, such as Bluetooth communications, "open" or in "discovery" mode (which allows the device to be seen by other Bluetooth-enabled devices so that connections can be made) could allow an attacker to install malware through that connection, or surreptitiously activate a microphone or camera to eavesdrop on the user. In addition, using unsecured public wireless Internet networks or WiFi spots could allow an attacker to connect to the device and view sensitive information.

In addition, connecting to an unsecured WiFi network could allow an attacker to access personal information from a device, putting users at risk for data and identity theft. One type of attack that exploits the WiFi network is known as man-in-the-middle, where an attacker inserts himself in the middle of the communication stream and steals information. For example, figure 4 depicts a man-in-the-middle attack using an unsecured WiFi network. As a result, an attacker within range could connect to a user's mobile device and access sensitive information.

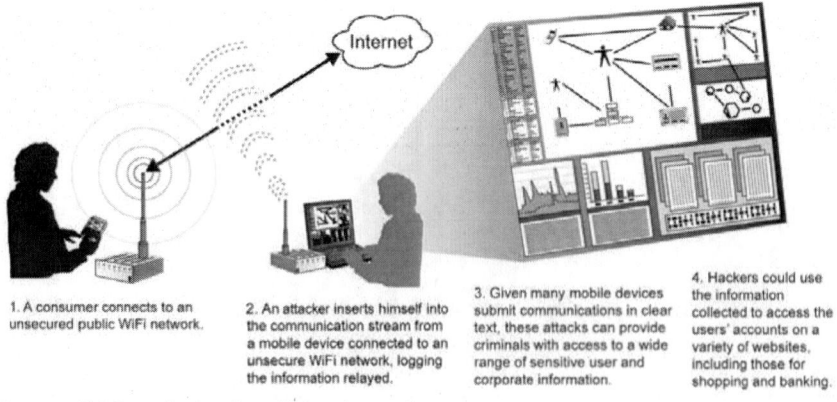

1. A consumer connects to an unsecured public WiFi network.
2. An attacker inserts himself into the communication stream from a mobile device connected to an unsecure WiFi network, logging the information relayed.
3. Given many mobile devices submit communications in clear text, these attacks can provide criminals with access to a wide range of sensitive user and corporate information.
4. Hackers could use the information collected to access the users' accounts on a variety of websites, including those for shopping and banking.

Source: GAO analysis of studies and security reports.

Figure 4. Man-in-the-Middle Attack Using an Unsecured WiFi Network.

| Security control | Description |
|---|---|
| Enable two-factor authentication for sensitive transactions | Two-factor authentication can be used when conducting sensitive transactions on mobile devices. Two-factor authentication provides a higher level of security than traditional passwords. Two-factor refers to an authentication system in which users are required to authenticate using at least two different "factors"—something you know, something you have, or something you are—before being granted access. Mobile devices themselves can be used as a second factor in some two-factor authentication schemes used for remote access. The mobile device can generate pass codes, or the codes can be sent via a text message to the phone. Two-factor authentication may be important when sensitive transactions occur, such as for mobile banking or conducting financial transactions. |
| Verify the authenticity of downloaded applications | Procedures can be implemented for assessing the digital signaturesaof downloaded applications to ensure that they have not been tampered with. |
| Install antimalware capability | Antimalware protection can be installed to protect against malicious applications, viruses,spyware, infected secure digital cards,band malware-based attacks. In addition, such capabilities can protect against unwanted (spam) voice messages, text messages, and e-mail attachments. |
| Install a firewall | A personal firewall can protect against unauthorized connections by intercepting both incoming and outgoing connection attempts and blocking or permitting them based on a list of rules. |
| Receive prompt security updates | Software updates can be automatically transferred from the manufacturer or carrier directly to a mobile device. Procedures can be implemented to ensure these updates are transmitted promptly. |
| Remotely disable lost or stolen devices | Remote disabling is a feature for lost or stolen devices that either locks the device or completely erases its contents remotely. Locked devices can be unlocked subsequentlyby the user if they are recovered. |
| Enable encryption for data stored on device or memory card | File encryption protects sensitive data stored on mobile devices and memory cards. Devices can have built-in encryption capabilities or use commercially available encryption tools. |
| Enable whitelisting | Whitelisting is a software control that permits only known safe applications to execute commands. |

Source: GAO analysis of guidance from NIST and DHS as well as recommended practices identified by the FCC CSRIC advisory committee.

[a] Digital signatures, or e-signature, are a way to communicate electronically and indicate that the person who claims to have written a message is the one who wrote it.

[b] A secure digital card is a memory card for use in portable devices.

Organizations may face different issues than individual consumers and thus may need to have more extensive security controls in place. For example, organizations may need additional security controls to protect proprietary and other confidential business data that could be stolen from mobile devices and need to ensure that mobile devices connected to the organization's network do not threaten the security of the network itself. Table 4 outlines controls that may be appropriate for organizations to implement to protect their networks, users, and mobile devices.

**Table 4. Additional Security Controls Specific to Organizations to Combat Common Threats and Vulnerabilities**

| Security control | Description |
|---|---|
| Adopt centralized security management | Centralized security management can ensure an organization's mobile devices are compliant with its security policies. Centralized security management includes (1) configuration control, such as installing remote disabling on all devices; and (2) management practices, such as setting policy for individual users or a class of users on specific devices. |
| Use mobile device integrity validation | Software tools can be used to scan devices for key compromising events (e.g., an unexpected change in the file structure) and then report the results of the scans, including a risk rating and recommended mitigation. |
| Implement a virtual private network (VPN) | A VPN can provide a secure communications channel for sensitive data transferred across multiple, public networks during remote access. VPNs are useful for wireless technologies because they provide a way to secure wireless local area networks, such as those at public WiFi spot, in homes, or other locations. |
| Use public key infrastructure (PKI) support | PKI-issued digital certificates can be used to digitally sign and encrypt e-mails. |
| Require conformance to government specifications | Organizations can require that devices meet government specifications before they are deployed. For example, NIST recommends that mobile devices used in government enterprises adhere to a minimum set of security requirements for cryptographic modules that include both hardware and software components. The Defense Information Systems Agency has certified a secure Android-based mobile system for use by DOD agencies. The system allows DOD personnel to sign, encrypt and decrypt e-mail, and securely access data from a smart phone or tablet computer. |
| Install an enterprise firewall | An enterprise firewall can be configured to isolate all unapproved traffic to and from wireless devices. |
| Monitor incoming traffic | Enterprise information technology network operators can use intrusion prevention software to examine traffic entering the network from mobile devices. |
| Monitor and control devices | Devices can be monitored and controlled for messaging, data leakage, inappropriate use, and to prevent applications from being installed. |
| Enable, obtain, and analyze device log files for compliance | Log files can be reviewed to detect suspicious activity and ensure compliance. |

Source: GAO analysis of guidance from NIST and DHS as well as recommended practices identified by the FCC CSRIC advisory committee.

[a] PKI is a system of hardware, software, policies, and people that, when fully and properly implemented, can provide a suite of information security assurances—including confidentiality, data integrity, authentication, and nonrepudiation—that are important in protecting sensitive communications and transactions.

## Security Practices for Mobile Devices

In addition to using mobile devices with security controls enabled, consumers can also adopt recommended security practices to mitigate threats and vulnerabilities. Table 5 outlines security practices consumers can adopt to protect the information on their devices. The practices are consistent with guidance from NIST and DHS, as well as recommended practices identified by FCC's CSRIC advisory committee.

### Table 5. Key Security Practices to Combat Common Threats and Vulnerabilities

| Security practice | Description |
|---|---|
| Turn off or set Bluetooth connection capabilities to nondiscoverable | When in discoverable mode, Bluetooth-enabled devices are "visible" to other nearby devices, which may alert an attacker to target them. When Bluetooth is turned off or in nondiscoverable mode, the Bluetooth-enabled devices are invisible to other unauthenticated devices. |
| Limit use of public WiFi networks when conducting sensitive transactions | Attackers may patrol public WiFi networks for unsecured devices or even create malicious WiFi spots designed to attack mobile phones. Public WiFi spots represent an easy channel for hackers to exploit. Users can limit their use of public WiFi networks by not conducting sensitive transactions when connected to them or if connecting to them, using secure, encrypted connections. This can help reduce the risk of attackers obtaining sensitive information such as passwords, bank account numbers, and credit card numbers. |
| Minimize installation of unnecessary applications | Once installed, applications may be able to access user content and device programminginterfaces, and they may also contain vulnerabilities. Users can reduce risk by limiting unnecessary applications. |
| Configure web accounts to use secure connections | Accounts for many websites can be configured to use secure, encrypted connections. Enabling this feature limits eavesdropping on web sessions. |
| Do not follow links sent in suspicious e-mail or text messages | Users should not follow links in suspicious e-mail or text messages, because such links may lead to malicious websites. |
| Limit clicking on suspicious advertisements within an application | Suspicious advertisements may include links to malicious websites, prompting the users to download malware, or violate their privacy. Users can limit this risk by not clicking on suspicious advertisements within applications. |
| Limit exposure of mobile phone numbers | By not posting mobile phone numbers to public websites, users may be able to limit the extent to which attackers can obtain known mobile numbers to attack. |
| Limit storage of sensitive information on mobile devices | Users can limit storing of sensitive information on mobile devices. |

**Table 5. (Continued)**

| Security practice | Description |
|---|---|
| Maintain physical control | Users can take steps to safeguard their mobile devices, such as by keeping their devices secured in a bag to reduce the risk that their mobile devices will be lost or stolen. |
| Delete all information stored in a device prior to discarding it | By using software tools that thoroughly delete (or "wipe") information stored in a device before discarding it, users can protect their information from unauthorized access. |
| Avoid modifying mobile devices | Modifying or "jailbreaking" mobile devices can expose them to security vulnerabilities or can prevent them from receiving security updates. |

Source: GAO analysis of guidance from NIST and DHS, as well as recommended practices identified by the FCC CSRIC advisory committee.

Organizations also benefit from establishing security practices for mobile device users. Table 6 outlines additional security practices organizations can take to safeguard mobile devices.

**Table 6. Additional Security Practices Specific to Organizations to Combat Common Threats and Vulnerabilities**

| Security practice | Description |
|---|---|
| Establish a mobile device security policy | Security policies define the rules, principles, and practices that determine how an organization treats mobile devices, whether they are issued by the organization or owned by individuals. Policies should cover areas such as roles and responsibilities, infrastructure security, device security, and security assessments. By establishing policies that address these areas, agencies can create a framework for applying practices, tools, and training to help support the security of wireless networks. |
| Provide mobile device security training | Training employees in an organization's mobile security policies can help to ensure that mobile devices are configured, operated, and used in a secure and appropriate manner. |
| Establish a deployment plan | Following a well-designed deployment plan helps to ensure that security objectives are met. |
| Perform risk assessments | Risk analysis identifies vulnerabilities and threats, enumerates potential attacks, assesses their likelihood of success, and estimates the potential damage from successful attacks on mobile devices. |
| Perform configuration control and management | Configuration management ensures that mobile devices are protected against the introduction of improper modifications before, during, and after deployment. |

Source: GAO analysis of guidance from NIST and DHS, as well as recommended practices identified by the FCC CSRIC advisory committee.

## PUBLIC AND PRIVATE-SECTOR ENTITIES HAVE TAKEN INITIAL STEPS TO ADDRESS SECURITY OF MOBILE DEVICES, BUT CONSUMERS REMAIN VULNERABLE TO THREATS

Federal agencies and mobile industry companies have taken steps to develop standards for mobile device security and have participated in initiatives to develop and implement certain types of security controls. However, these efforts have been limited in scope, and mobile device manufacturers and carriers do not consistently implement security safeguards on mobile devices. Although FCC has facilitated public-private coordination to address specific challenges, such as cellphone theft, and developed cybersecurity best practices, it has not yet taken similar steps to encourage device manufacturers and wireless carriers to implement a more complete industry baseline of mobile security safeguards. Furthermore, DHS, FTC, NIST, and the private sector have taken steps to raise public awareness about mobile security threats. However, security experts agree that many consumers still do not know how to protect themselves from mobile security vulnerabilities. DHS and NIST have not yet developed performance measures that would allow them to determine whether they are making progress in improving awareness of mobile security issues.

## EFFORTS HAVE BEEN MADE TO ADDRESS SECURITY VULNERABILITIES, BUT CONTROLS ARE NOT ALWAYS IMPLEMENTED

Federal agencies and mobile industry companies have worked to develop best practices and taken steps to address certain aspects of mobile security.

FCC has worked with mobile companies on several initiatives. For example, FCC tasked its advisory committee, CSRIC, with developing cybersecurity best practices, including recommended practices for wireless and mobile security. In March 2011, CSRIC released its report recommending that wireless carriers and device manufacturers consider adopting practices such as:[29]

- working closely and regularly with customers to provide recommendations concerning existing default settings and to identify future default settings that may introduce vulnerabilities;
- employing fraud detection systems to detect customer calling anomalies (e.g., system access from a single user from widely dispersed geographic areas);
- having processes in place to ensure that all third-party software has been properly patched with the latest security patches and that the system works correctly with those patches installed;
- establishing application support for cryptography that is based on open and widely reviewed and implemented encryption algorithms and protocols; and
- enforcing strong passwords for mobile device access and network access.

In addition, in March 2012 FCC tasked CSRIC with examining three major cybersecurity threats to networks that allow cyber criminals to access Internet traffic for theft of personal information and intellectual property. In response, CSRIC recommended that wireless carriers (1) use key practices when mitigating botnet threats, (2) use best practices for deploying and managing Domain Name System Security Extensions,[30] and (3) develop an industry framework to prevent Internet route hijacking via security weaknesses in the Border Gateway Protocol.[31] CSRIC is working with the wireless carriers to implement these recommendations and is tasked with developing ways to measure the effectiveness of the recommendations.

FCC also tasked the Technological Advisory Council's Wireless Security and Privacy working group to examine mobile security issues, such as vulnerabilities of WiFi networks, security of older generation cellular networks, malicious applications, and text messaging security. The working group is scheduled to issue its recommendations in December 2012.

Moreover, in April 2012, FCC announced that it had reached agreement with the CTIA-the Wireless Association,[32] and multiple wireless carriers to establish processes to deter theft of mobile devices. Under the antitheft agreement, participating wireless carriers are to take several specific actions and submit quarterly progress reports to FCC. For example, the antitheft agreement calls for wireless carriers to initiate, implement, and deploy database solutions by October 31, 2012, to prevent reportedly lost or stolen smartphones from being used on another wireless network. The FCC plans to monitor progress in developing these databases and CTIA agreed to report progress quarterly, beginning June 30, 2012. The agreement also will result in the launch of a public education campaign by July 1, 2012, to inform consumers about the ability to lock or locate and erase data from a smartphone.

In addition, wireless carriers and device manufacturers reported that they participate in private-sector standards-setting organizations, which have addressed aspects of mobile security. For example, the Open Mobile Alliance, an industry standards group, has developed a specification to provide a common means for mobile developers to implement standards for secure and reliable data transport between two communicating parties. Furthermore, a consortium of wireless carriers and mobile device manufacturers known as the Messaging, Malware and Mobile Anti-Abuse Working Group 33 has an initiative underway to address text-messagebased spam. Under this initiative, wireless carriers encourage customers to forward spam text messages back to the carriers, who can use the messages to identify the source of spam and take corrective action to block its content from their networks. According to FCC officials, the current chairman of the Messaging, Malware and Mobile Anti-Abuse Working Group is a member of CSRIC and the chair of the working group that developed recommended solutions for the botnet threats.

## Despite Efforts, Mobile Security Safeguards Are Not Always Implemented

While private and public sector entities have initiated activities to identify mobile security safeguards, these safeguards are not always available on mobile devices or activated by users. According to a 2012 study by NQ Mobile and the National Cyber Security Alliance (NCSA),[34] approximately 30 percent of respondents[35] said they did not have mobile security features on their smartphones.[36] In addition, approximately 66 percent of respondents did not report that they activated password protection on their devices to prevent unauthorized access and that at least 67 percent did not report activating a remote-wipe or remote-locate security feature. Security company representatives told us that these results were generally consistent with their experiences and observations.

In addition, mobile device manufacturers and wireless carriers do not consistently implement or activate security safeguards on their mobile devices. According to most of the device manufacturers and several wireless carriers we spoke with, safeguards such as passwords, encryption, and remote wipe/lock/locate can be made available on their mobile devices, although one wireless carrier noted that encryption might be inappropriate for certain types of devices. Several of these companies also acknowledged that it is possible to preconfigure mobile devices to prompt the user to implement safeguards when the phone is first set up. However, with the exception of password protection for online voicemail accounts, none of the device manufacturers or wireless carriers stated that they generally configure their devices to prompt the user to implement these controls. We also observed that general cybersecurity instructions were not directly accessible from either carriers or device manufacturers, although instructions for implementing controls could be found by searching the company's website for information about individual models of smartphones.

## FCC Has Facilitated Industry Best Practice Efforts but Has Not Yet Encouraged Broad Implementation of Mobile Security Safeguards

FCC has the ability to encourage broad implementation of mobile security safeguards among mobile industry companies. While it has taken steps to encourage implementation of safeguards in certain areas, it has not yet taken similar steps to encourage industry implementation of a broad baseline of mobile security safeguards. For example, in its recent antitheft agreement with CTIA, and participating wireless carriers, FCC took an active role in encouraging major wireless carriers to adopt specific procedures to discourage the theft of mobile devices. This effort demonstrates that FCC can facilitate private sector efforts to establish an industry baseline and milestones for addressing mobile security challenges. Moreover, representatives from multiple companies agreed that FCC could play a role in coordinating private sector efforts to improve mobile security.

FCC has also facilitated private sector efforts to establish cybersecurity best practices in areas not specific to mobile security. As mentioned previously, FCC tasked CSRIC to review best practices for botnet threats, Domain Name System attacks, and Internet route hijacking. CSRIC developed voluntary recommendations in these areas and has been working with wireless carriers to implement them. According to FCC officials, wireless carriers representing 90 percent of the domestic customer base have committed to adopting and using these practices. Although these recommendations are not specific to mobile devices, FCC officials stated that the process of seeking voluntary compliance from carriers had been successful and demonstrated the willingness of carriers to adopt best practices.

FCC officials stated that they hope to have the same cooperation from wireless carriers when the Technological Advisory Council's Wireless Security and Privacy working group releases its recommendations on mobile security issues, scheduled for December 2012. While it is not clear that the working group will develop a baseline of recommended practices for implementation by mobile industry companies, the council's recommendations nevertheless could be part of such a baseline.

Another candidate for a set of baseline mobile security standards that mobile industry companies could be encouraged to implement is the collection of cybersecurity best practices developed by CSRIC in 2011. Those practices have not yet been adopted as a baseline within the mobile industry. FCC officials from the Public Safety and Homeland Security Bureau stated that they had not yet taken action to promote this specific set of recommended practices, although they had held informal meetings with industry to discuss the implementation of cybersecurity practices. Whether mobile industry companies adopt the CSRICrecommended practices or choose other baseline practices and controls, it will be important for FCC to encourage industry to adopt recommended practices. If such practices are not implemented, vulnerabilities in mobile devices are likely to continue to pose risks for consumers.

## The Effectiveness of Public and Private Sector Efforts to Raise Awareness about Mobile Security Is Unclear

Many of the key practices that have been identified as effective in mitigating mobile security risks depend on the active participation of users. Thus it is important that an appropriate level of awareness is achieved among consumers who use mobile devices on a regular basis. To address this need, federal agencies have developed and distributed a variety of educational materials. For example:

- DHS's US-Computer Emergency Readiness Team (US-CERT) has developed cybersecurity tip sheets and technical papers related to mobile security. These materials, which are published on the US-CERT website, provide lists of suggestions, such as the use of

- passwords and encryption, to help consumers to protect their devices and sensitive data from network attacks and theft.[37]
- DHS coordinates domestic and international engagements and cybersecurity outreach endeavors. For example, as the lead agency for the awareness component of the NICE initiative, DHS coordinates the National Cyber Security Awareness Month and a national cybersecurity public awareness campaign called "Stop. Think. Connect." As part of these efforts, DHS has developed educational materials that, although not specifically related to mobile security, encourage users to adopt safe practices when using the Internet. The DHS website related to this effort also provides links to educational materials hosted on third-party websites, such as StaySafeOnline.org.
- FTC manages the OnGuardOnline website,[38] which provides individuals with information about how to use the Internet in a safe, secure, and responsible manner. As part of this effort, FTC has developed educational materials specifically related to mobile security, such as avoiding malicious mobile applications and protecting children who use mobile devices.[39] In addition, FTC and DHS have developed and distributed printed cybersecurity guides to schools, business, and other entities, according to an FTC staff member.
- NIST published guidelines on the security of cellphones and personal digital assistants in 2008. [40] Among other things, this guidance provides users with information about how to secure their devices. For example, the guidance discusses the value of implementing authentication (e.g., password protection) and remotely erasing or locking devices that are lost or stolen.

DHS and nonprofit organizations also have developed and distributed cybersecurity educational materials in collaboration with NCSA. In addition to funding from the private sector, DHS officials stated that DHS has contributed a grant to NCSA to conduct surveys and other activities. NCSA has produced educational materials that specifically relate to mobile security. For example, NCSA's website provides tips that individuals can follow to protect their mobile devices such as avoiding malware, using trusted internet connections, and securing personal information through the use of strong passwords.

Other private sector organizations have also developed educational materials related to securing mobile devices. For example, the Global System for Mobile Communications Association has published articles targeted towards mobile phone users on topics, such as (1) preventing mobile phone theft, (2) spam and mobile phones, and (3) computer viruses and mobile phones.[41] Similarly, CTIA maintains a blog with information on topics such as establishing passwords and using applications that can track, locate, lock, and/or wipe wireless devices that are lost or stolen. In addition, as part of the antitheft initiative discussed above, CTIA agreed that its members would implement a system to inform users about security safeguards on mobile devices as well as launch an education campaign regarding the safe use of smartphones.

Despite the efforts underway by the federal government and the private sector to develop and distribute educational materials, it is unclear whether consumer awareness has improved as a result. Representatives from companies that specialize in information security told us that many consumers do not understand the importance of implementing mobile security safeguards or do not know how to implement them. Their views are consistent with the results of the 2012 NCSA study, which suggested that many mobile users do not know how to implement mobile security safeguards. The survey reported that more than half of respondents felt that they required additional information in order to select and/or implement security solutions for their mobile devices. Further, the study reported that approximately three-quarters of respondents reported that they did not receive information about the need for security solutions at the time they purchased their phone. The survey did not include data that would indicate whether consumer awareness had improved or worsened over time.

## DHS and NIST Have Not Determined Whether Efforts Are Improving Consumer Awareness

While DHS and NIST have conducted or supported several consumer cybersecurity awareness efforts, neither has developed outcome-oriented performance measures to assess the effectiveness of government efforts to enhance consumer awareness of mobile security. An outcome-oriented performance measure is an assessment of the result, effect, or consequence that will occur from carrying out a program or activity compared to its intended purpose. NIST officials stated that they do not currently measure progress related to awareness activities associated with NICE. Furthermore, although DHS officials stated that the department assesses the effectiveness of several of the awareness activities, these assessments are not based on outcome-oriented measures. For example, DHS officials stated that they assess the "Stop. Think. Connect." events by (1) the number of individuals who join the campaign and agree to receive additional information, such as newsletters, concerning cybersecurity; (2) the total number of events held; (3) the number of agencies and states that join the campaign; and (4) the number of times the campaign website is visited. However, these measures are not outcome-oriented because they do not indicate how, if at all, these activities have (1) improved citizens' knowledge about managing online risk, (2) improved knowledge of cybersecurity within organizations, or (3) enabled access to cybersecurity resources.

To develop measures of the impact of government efforts on consumer awareness of mobile security issues, a baseline measure of consumer awareness would be needed from which to mark progress. However, neither DHS nor NIST has developed a baseline measure of the state of national cybersecurity awareness. Establishing a baseline measure and regularly assessing consumer awareness and behavior regarding a particular issue can enable organizations to document where problems exist, identify causes, prioritize efforts, and monitor progress. DHS officials stated that the department has considered conducting a study on consumer behavior and awareness related to general cybersecurity but has not yet done so.

Without a baseline measure of consumer awareness, it will remain difficult for NIST and DHS to measure any correlation between the government's activities and enhanced consumer awareness. Further, without outcome-oriented performance measures, the government will be limited in its ability to determine whether it is achieving its identified goals and objectives, including whether cybersecurity awareness efforts are effective at increasing adoption of recommended security practices.

## CONCLUSION

Mobile devices face an array of threats that take advantage of numerous vulnerabilities commonly found in such devices. These vulnerabilities can be the result of inadequate technical controls, but they can also result from the poor security practices of consumers.

Private sector entities and relevant federal agencies have taken steps to improve the security of mobile devices, including making certain controls available for consumers to use if they wish and promulgating information about recommended mobile security practices. However, security controls are not always consistently implemented on mobile devices, and it is unclear whether consumers are aware of the importance of enabling security controls on their devices and adopting recommended practices.

Although FCC has taken steps to work with industry to develop cybersecurity best practices, it has not yet taken steps to encourage wireless carriers and device manufacturers to implement a more complete industry baseline of mobile security safeguards, and NIST and DHS have not determined whether consumer awareness of mobile security issues has improved since the government's efforts have been initiated.

## RECOMMENDATIONS FOR EXECUTIVE ACTION

To help mitigate vulnerabilities in mobile devices, we recommend that the Chairman of the Federal Communications Commission

- continue to work with wireless carriers and device manufacturers on implementing cybersecurity best practices by encouraging them to implement a complete industry baseline of mobile security safeguards based on commonly accepted security features and practices; and
- monitor progress of wireless carriers and device manufacturers in achieving their milestones and time frames once an industry baseline of mobile security safeguards has been implemented.

To determine whether the NICE initiative is having a beneficial effect in enhancing consumer awareness of mobile security issues, we recommend that the Secretary of Homeland Security in collaboration with the Secretary of Commerce

- establish a baseline measure of consumer awareness and behavior related to mobile security and
- develop performance measures that use the awareness baseline to assess the effectiveness of the awareness component of the NICE initiative.

## AGENCY COMMENTS AND OUR EVALUATION

We received written comments on a draft of this report from the Chief of FCC's Public Safety and Homeland Security Bureau, the Director of DHS's Departmental GAO-OIG Liaison Office, and the Acting Secretary of Commerce. These officials generally concurred with our recommendations and provided technical comments, which we have considered and incorporated as appropriate into the final report. FTC did not provide written comments on the draft report, but an attorney in FTC's Office of the General Counsel did provide technical comments in an e-mail that we addressed as appropriate. DOD did not provide comments on the draft report. The comments we received are summarized below.

- In addition to FCC's written comments, the Chief of FCC's Public Safety and Homeland Security Bureau stated in e-mail comments that the commission generally concurred with our recommendations that it encourage wireless carriers and device manufacturers to implement a complete industry baseline of mobile security safeguards; and to monitor progress of wireless carriers and device manufacturers in achieving their milestones and time frames once a baseline has been implemented. In the written comments, the Chief added that FCC has facilitated private sector efforts, for example, through advisory committees such as CSRIC to establish and promote the implementation of cybersecurity best practices that secure the underlying Internet infrastructure. FCC officials also provided preliminary oral and written technical comments, which we addressed as appropriate.
- The Director of DHS's Departmental GAO-OIG Liaison Office provided written comments in which the department concurred with our recommendations that it work with Commerce to establish a baseline measure of consumer awareness and behavior related to mobile security and develop performance measures that use the baseline to assess the effectiveness of the awareness component of the NICE initiative. He stated that the department will coordinate with its counterparts at Commerce to assess the feasibility of different methods to create a baseline measure of consumer awareness and continue to promote initiatives to educate the public about cybersecurity. He also stated that the department will coordinate with its NIST counterparts on the development of performance measures

- using the awareness campaign and other methods. He also provided technical comments, which we have incorporated as appropriate.
- The Acting Secretary of Commerce provided written comments in which the department concurred in principle with our recommendations that NIST work with DHS to establish a baseline measure of consumer awareness and behavior related to mobile security and that it develop performance measures that use the baseline to assess the effectiveness of the awareness component of the NICE initiative. The Acting Secretary provided technical comments and asked that we consider replacing "baseline understanding" with "baseline measure," which we have incorporated into the final report. She also provided suggested revised text. However, we believe that the information in the draft is correct and communicates appropriately as written. Therefore, we have not added the suggested text.

Gregory C. Wilshusen
Director
Information Security Issues

Dr. Nabajyoti Barkakati
Chief Technologist

# APPENDIX I: OBJECTIVES, SCOPE, AND METHODOLOGY

The objectives of our review were to determine: (1) what common security threats and vulnerabilities currently exist in mobile devices, (2) what security features are currently available and what practices have been identified to mitigate the risks associated with these vulnerabilities, and (3) the extent to which government and private entities are addressing security vulnerabilities of mobile devices.

To determine the common security threats and vulnerabilities that currently exist in mobile devices (cellphones, smartphones, and tablets), as well as security features and practices to mitigate them, we identified agencies and private companies with responsibilities in the telecommunication and cybersecurity arena, and reviewed and analyzed information security-related websites, white papers, and mobile security studies. We interviewed officials, and obtained and analyzed documentation from the Federal Communications Commission (FCC), Department of Homeland Security (DHS), Department of

Defense (DOD), Department of Commerce (Commerce), and Federal Trade Commission (FTC) to determine the extent to which they have identified mobile security vulnerabilities and developed standards and guidance on the security of mobile devices. We interviewed and obtained documents from an industry group and an advisory council, both of which have representation from the telecommunication industry; these included the CTIA-The Wireless Association and the Communications Security, Reliability, and Interoperability Council (CSRIC). We also analyzed information from the US-Computer Emergency Readiness Team (USCERT) and the National Vulnerability Database on mobile security vulnerabilities.

Further, we obtained input from the private companies who make up the largest market share for mobile devices in the United States to determine what steps they are taking to provide security for their mobile devices. These included mobile device manufacturers—HTC Corporation, Research In Motion, Corp, Motorola Mobility Inc., Samsung, and LG Electronics—as well as wireless carriers—Verizon Wireless, AT&T Inc., T-Mobile USA Inc., and Sprint. We also met with representatives of information security companies, including Symantec Corporation and Juniper Networks. We approached Apple Inc. and Google Inc.; however, Apple officials did not agree to meet with us and Google officials did not provide responses to our questions. We developed draft lists of common vulnerabilities and security practices based on our analysis of government security guidance as well as private sector studies and reports. We provided copies of these lists to each of the companies listed above and addressed their comments as appropriate.

To determine the extent to which government and private entities are addressing security vulnerabilities of mobile devices, we analyzed statutes and regulations to determine federal roles related to mobile security. In order to identify initiatives related to improving mobile security or raising consumer awareness, we interviewed the federal and private sector officials mentioned above, and members of a private sector working group devoted to mobile security issues, known as the Messaging, Malware, and Mobile Anti-Abuse Working Group. In addition, we analyzed multiple studies related to consumer attitudes and practices related to mobile devices. Specifically, we assessed available methodological information against general criteria for survey quality and relevant principles derived from the Office of Management and Budget (OMB) Standards and Guidelines for Statistical Surveys. Because the available methodological documentation did not allow us to fully assess the quality of the survey data, the risk of error in the surveys makes it possible that reported results may not be very accurate or precise. Although we corroborated the study's general findings with information security experts, readers should be cautious in drawing conclusions based on these results.

We conducted this performance audit from November 2011 to September 2012 in accordance with generally accepted government auditing standards. Those standards require that we plan and perform the audit to obtain sufficient, appropriate evidence to provide a reasonable basis for our findings and conclusions based on our audit objectives. We believe that the evidence obtained provides a reasonable basis for our findings and conclusions based on our audit objectives.

## APPENDIX II: FEDERAL WEBSITES FOR INFORMATION RELATED TO MOBILE SECURITY

The table below provides information and website links to federal sites that include information related to mobile security. Website links are current as of July 10, 2012.

**Table 7. Federal Websites and Links to Information Related to Mobile Security**

| Agency | Information related to mobile security | Links |
|---|---|---|
| DOD | DOD maintains a website on Security Technical Implementation Guides, which contain technical guidance to secure information systems or software that might otherwise be vulnerable to a malicious computer attack. In addition, the guides address aspects of mobile device security. | http://iase.disa.mil/stigs/net_perimeter/wireless/smartphone.html |
| DHS | DHS maintains a website called Cybersecurity Tips that provides general cybersecurity tips as well as a section specific to mobile devices. DHS also launched a website to promote the "Stop.Think.Connect." initiative. This website provides information to visitors on the initiative itself, top security issues, cybersecurity tips, and links to additional resources.<br>US-CERT also maintains a website with cybersecurity tips. Information on threats to the security of mobile devices is available in addition to more general cybersecurity information. | http://www.dhs.gov/files/events/cybersecurity<br>http://www.dhs.gov/files/events/stop-think-connect.shtm<br>http://www.us-cert.gov/cas/tips<br>http://www.us-cert.gov/cas/tips/ST04-017.html<br>http://www.us-cert.gov/cas/tips/ST04-020.html |

## Table 7. (Continued)

| Agency | Information related to mobile security | Links |
|---|---|---|
| FCC | FCC maintains a website with public safety tech topics that provides detailed information on the threats and vulnerabilities associated with the use of various communication technologies, including mobile devices. | http://www.fcc.gov/guides/stolen-and-lost-wireless-devices<br>http://www.fcc.gov/help/topic/106<br>http://www.fcc.gov/help/public-safety<br>http://transition.fcc.gov/pshs/techtopics/ |
| FTC | FTC maintains a list of consumer publications as well as links to information on information security. The website "OnGuard Online" provides information for consumers to protect their devices while on the Internet. In addition, the FTC website has tips on how to properly dispose of mobile devices. | http://www.ftc.gov/bcp/edu/microsites/idtheft/consumers<br>http://www.ftc.gov/bcp/menus/consumer/data/privacy.shtm<br>http://www.ftc.gov/bcp/edu/pubs/consumer/alerts/alt044.shtm |
| NIST | NIST's homepage has a publications link that directs customers to a web page containing a publications search engine. This search engine enables customers to search through a database of publications, related to cybersecurity, maintained by NIST. One of the publications located within this database details information about the threats and technology risks associated with the use of mobile devices and available safeguards to mitigate them (Guidelines on Cell Phone and PDA Security Special Publication 800-124). | http://csrc.nist.gov/groups/SNS<br>http://csrc.nist.gov/publications/nistpubs/800-124/SP800-124.pdf |

Source: GAO analysis of federal websites related to mobile security.

## End Notes

[1] Lookout Mobile Security, *Lookout Mobile Threat Report* (San Francisco, Calif.: August 2011).

[2] Pew Research Center, *46% of American Adults Are Smartphone Owners* (Washington, D.C.: March 2012).

[3] WiFi and Bluetooth are commonly used technologies that allow an electronic device to exchange data wirelessly (using radio waves) with other devices and computer networks.

[4] Pew Research Center, *46% of American Adults Are Smartphone Owners* (Washington, D.C.: March 2012).

[5] Pew Research Center, *Tablet and E-book Reader Ownership Nearly Double Over the Holiday Gift-Giving Period* (Washington, D.C.: January 2012).

[6] The Open Handset Alliance is a consortium of 84 hardware, software, and telecommunications companies devoted to advancing open standards for mobile devices.

[7] Java is a programming language and computing platform that powers programs including utilities, games, and business applications.

[8] The Institute of Electrical and Electronics Engineers is a professional association focused on electrical and computer sciences, engineering, and related disciplines. It is responsible for developing technical standards through its Standards Association, which follows consensus-based standards development processes.

[9] Bluetooth is an open standard for short-range radio frequency communication. Bluetooth technology is used primarily to establish wireless personal area networks, commonly referred to as ad hoc or peer-to-peer networks. The standard allows mobile devices to be placed in different modes: discoverable, which allows the device to be detected and receive connections from other Bluetooth-enabled devices; connectable, which allows the device to respond to other devices and establish a network connection with them; or completely off.

[10] Homeland Security Presidential Directive 7 establishes a national policy for federal departments and agencies to identify and prioritize critical infrastructure and to protect them from terrorist attacks. The directive defines relevant terms and delivers 31 policy statements. These policy statements define what the directive covers and the roles various federal, state, and local agencies will play in carrying it out.

[11] 15 U.S.C. 278g-3, as amended by FISMA, Title III, Pub. L. No. 107-347 (Dec. 17, 2002).

[12] NIST, *Guidelines on Cell Phone and PDA Security, SP 800-124* (Gaithersburg, Md.: October 2008) and *Guide to Bluetooth Security, SP 800-121, Revision 1* (Gaithersburg, Md.: June 2012).

[13] GAO, *Cybersecurity Human Capital: Initiatives Need Better Planning and Coordination*, GAO-12-8 (Washington, D.C.: Nov. 29, 2011).

[14] The White House, *The National Strategy to Secure Cyberspace* (Washington, D.C.: February 2003).

[15] See for example, 47 U.S.C. 151 and 332; Communications Act of 1934, as amended, including by the Telecommunications Act of 1996, Pub. L. No. 104-104 (Feb. 8, 1996).

[16] The CSRIC has operated under 2-year charters that have regularly been renewed.

[17] 15 U.S.C. 45.

[18] OMB, *Memorandum for the Heads of Departments and Agencies: Protection of Sensitive Agency Information M-06-16* (Washington, D.C.: June 23, 2006).

[19] Threats are any circumstance or event with the potential to adversely impact organizational operations (including mission, functions, image, or reputation), organizational assets, individuals, other organizations, or the nation through an information system via unauthorized access, destruction, disclosure, modification of information, and/or denial of service.

[20] Underscoring the importance of this issue, we have designated federal information security as a high-risk area since 1997. See, most recently, GAO, *High-Risk Series: An Update*, GAO-11-278 (Washington, D.C.: February 2011).

[21] Vulnerabilities are weaknesses in an information system, system security procedures, internal controls, or implementation that could be exploited or triggered by a threat source.

[22] A botnet is a collection of compromised systems, each of which is known as a 'bot,' connected to the Internet. When a mobile device is compromised by an attacker, there is often code within the malware that commands it to become part of a botnet. The botnet's operator remotely controls these compromised mobile devices.

[23] Juniper Networks, Inc., *2011 Mobile Threats Report* (Sunnyvale, Calif.: February 2012).

[24] Symantec Corporation, *Internet Security Threat Report, 2011 Trends* Vol.17 (Mountain View, Calif.: April 2012).

[25] Lookout Mobile Security, *Lookout Mobile Threat Report* (San Francisco, Calif.: August 2011).

[26] Lookout Mobile Security, *Lookout Mobile Threat Report* (San Francisco, Calif.: August 2011).

[27] Http is an application protocol that allows the transmitting and receiving of information across the Internet. While http allows for the quick transmission of information it is not secure and it is possible for a third party to intercept the communication. The secure http protocol encrypts http and was developed to allow the authorization of users and secure transactions.

[28] Juniper Networks, Inc., *2011 Mobile Threats Report* (Sunnyvale, Calif.: February 2012), Symantec Corporation, *Internet Security Threat Report, 2011 Trends Vol.17* (Mountain View, Calif.: April 2012), Lookout Mobile Security, *Lookout Mobile Threat Report* (San Francisco, Calif.: August 2011), McAfee, *Securing Mobile Devices: Present and Future* (Santa Clara, Calif.: 2011).

[29] CSRIC, *Working Group 2A Cyber Security Best Practices, Final Report* (Washington, D.C.: March 2011). The CSRIC Working Group's security best practices are mostly technical in nature and the examples provided are high-level examples of wireless and mobile security practices. A copy of the best practices can be obtained from the CSRIC website, date accessed March 13, 2012, http://transition.fcc.gov/pshs/advisory/csric/.

[30] The Domain Name System converts domain names to numerical IP addresses. Security shortcomings in the Domain Name System have enabled spoofing, allowing Internet criminals to steal credit card numbers and personal data from users who do not realize they have been sent to an illegitimate website. Domain Name System Security Extensions have been developed to prevent such fraudulent activity.

[31] The Border Gateway Protocol is the protocol that allows seamless connectivity among the networks that make up the Internet. It does not have built-in security measures. Thus, Internet traffic can be misdirected through potentially untrustworthy networks such as those operated by cyber criminals or by foreign governments.

[32] CTIA is an international nonprofit membership organization that has represented the wireless communications industry since 1984. Membership in the association includes wireless carriers and their suppliers, as well as providers and manufacturers of wireless data services and products.

[33] The Messaging, Malware and Mobile Anti-Abuse Working Group, whose members include wireless carriers and handset manufacturers, among others, is a private organization that collaborates to address online challenges such as web messaging abuse and botnets.

[34] The NCSA is a nonprofit organization whose mission is to promote secure and safe use of the Internet. NCSA leadership includes several private companies, including two major wireless carriers.

[35] NQ Mobile and National Cyber Security Alliance, *Report on Consumer Behaviors and Perceptions of Mobile Security* (Jan. 25, 2012), date accessed April 10, 2012, http://docs.nq.com/NQ_Mobile_Security_Survey_Jan2012.pdf.

[36] Participants in this web survey had Internet access and were recruited from visitors to selected websites and other sources. The survey is not based on a random probability sample and is not necessarily representative of the larger population of cellphone users in the United States. See app. I for additional information about this survey.

[37] Examples of US-CERT tip sheets include *Protecting Portable Devices: Data Security*, date accessed March 19, 2012, http://www.us-cert.gov/cas/tips/ST04-020.html; and *Protecting Portable Devices: Physical Security*, date accessed July 3, 2012, http://www.us-cert.gov/cas/tips/ST04-017.html.

[38] The OnGuardOnline website can be accessed at http://onguardonline.gov/, date accessed August 9, 2012.

[39] *Understanding Mobile Apps* (September 2011), date accessed June 14, 2012, http://onguardonline.gov/articles/0018-understanding-mobile-apps; *Kids and Mobile Phones* (September 2011), date accessed June 14, 2012, http://onguardonline.gov/articles/0025-kids-and-mobile-phones.

[40] NIST, *Special Publication 800-124, Guidelines on Cell Phone and PDA Security* (October 2008). According to NIST officials, they are revising this publication and will release a draft update in fiscal year 2012.

[41] GSMA, *Security Advice for Mobile Phone Users,* date accessed June 14, 2012, http://www.gsma.com/security

In: Mobile Device Security
Editor: Willliam R. O'Connor

ISBN: 978-1-62417-254-0
© 2013 Nova Science Publishers, Inc.

*Chapter 2*

# TECHNICAL INFORMATION PAPER: CYBER THREATS TO MOBILE DEVICES[*]

## *United States Computer Emergency Readiness Team*

### OVERVIEW

Today's advanced mobile devices are well integrated with the Internet and have far more functionality than mobile phones of the past. They are increasingly used in the same way as personal computers (PCs), potentially making them susceptible to similar threats affecting PCs connected to the Internet. Since mobile devices can contain vast amounts of sensitive and personal information, they are attractive targets that provide unique opportunities for criminals intent on exploiting them. Both individuals and society as a whole can suffer serious consequences if these devices are compromised.

This paper introduces emerging threats likely to have a significant impact on mobile devices and their users.

---

[*] This is a reformatted and augmented version of a US-CERT Technical Information Paper - TIP-10-105-01 released on April 15, 2010.

## INTRODUCTION

As mobile device technology evolves, consumers are using it at unprecedented levels. Mobile cellular technology has been the most rapidly adopted technology in history, with an estimated 4.6 billion mobile cellular subscriptions globally at the end of 2009.[1] Furthermore, technological advances have fueled an unprecedented portable computing capability, increasing user dependence on mobile devices and skyrocketing mobile broadband subscriptions. Mobile broadband connections rose by more than 850% in 2008,[2] exceeding the number of fixed broadband subscribers.[3] Mobile devices have become an integral part of society and, for some, an essential tool. However, the complex design and enhanced functionality of these devices introduce additional vulnerabilities. These vulnerabilities, coupled with the expanding market share, make mobile technology an attractive, viable, and rewarding target for those interested in exploiting it.

In the past, malicious activity targeting mobile phones was relatively limited compared to that of PCs. The proprietary nature and limited functionality of the hardware and software architectures previously used by individual mobile phone manufacturers made this market a less than ideal target for mass exploitation. Current mobile devices have much greater functionality and more accessible architectures, resulting in an increase in malicious activity affecting them. These smartphones include the Apple iPhone, Google Android, Research in Motion (RIM) Blackberry, Symbian, and Windows Mobile-based devices.

Due to the similar functionality of mobile devices and PCs, the distinction between the two has blurred. Mobile devices have become equally susceptible to malicious cyber activity and will likely be affected by many of the same threats that exist for PCs on the Internet. The variety of sensitive information available from a mobile device is also potentially greater and more enticing than that of a traditional mobile phone or computer. Users are more likely to take advantage of the portability and convenience of mobile devices for activities such as banking, social networking, emailing, and maintaining calendars and contacts. The features of mobile devices also introduce additional types of information not typically available from a PC, such as information related to global positioning system (GPS) functionality and text messaging.

A multitude of threats exist for mobile devices, and the list will continue to grow as new vulnerabilities draw the attention of malicious actors. This paper provides a brief overview of mobile device malware and provides information on the following threats to mobile devices:

- Social engineering;
- Exploitation of social networking;
- Mobile botnets;
- Exploitation of mobile applications; and
- Exploitation of m-commerce.

## MOBILE MALWARE

Malicious actors have created and used malware targeted to mobile devices since at least 2000. The total number of malware variants significantly increased in 2004 with the public release of Cabir source code.[4] Cabir is a Bluetooth worm and the first widespread sample of mobile malware. It runs on mobile phones using the Symbian Series 60 platform and spreads among Bluetooth-enabled devices that are in discoverable mode. The worm causes a phone to constantly attempt to make a Bluetooth connection, subsequently draining the battery. While this worm was an inconvenience to device users, today's mobile malware is more insidious and often has more severe effects on devices and their users.

A recent and more nefarious example of mobile malware is the Ikee.B, the first iPhone worm created with distinct financial motivation. It searches for and forwards financially sensitive information stored on iPhones and attempts to coordinate the infected iPhones via a botnet command and control server.[5] This worm only infects iPhones that have a secure shell (SSH) application installed to allow remote access to the device, have the root password configured as "alpine"—the factory default—and are "jailbroken." A jailbroken iPhone is one that has been configured to allow users to install applications that are not officially distributed by Apple. Although Ikee.B has limited growth potential, it provides a proof of concept that hackers can migrate the functionality typical to PC-based botnets to mobile devices. For example, a victim iPhone in Australia can be hacked from another iPhone located in Hungary and forced to exfiltrate its user's private data to a Lithuanian command and control server.[6]

Spy software also exists for mobile devices, including some programs being sold as legitimate consumer products. FlexiSpy is commercial spyware sold for up to $349.00 per year. Versions are available that work on most of the major smartphones, including Blackberry, Windows Mobile, iPhone, and Symbian-based devices. The following are some of the capabilities provided by the software:[7]

- Listen to actual phone calls as they happen;
- Secretly read Short Message Service (SMS) texts, call logs, and emails;
- Listen to the phone surroundings (use as remote bugging device);
- View phone GPS location;
- Forward all email events to another inbox;
- Remotely control all phone functions via SMS;
- Accept or reject communication based on predetermined lists; and
- Evade detection during operation.

FlexiSpy claims to help protect children and catch cheating spouses, but the implications of this type of software are far more serious. Imagine a stranger listening to every conversation, viewing every email and text message sent and received, or tracking an individual's every movement without his or her knowledge. FlexiSpy requires physical access to a target phone for installation; however, these same capabilities could be maliciously exploited by malware unknowingly installed by a mobile user.

Cross-platform mobile malware further complicates the issue. The Cardtrp worm infects mobile devices running the Symbian 60 operating system and spreads via Bluetooth and Multimedia Messaging Service (MMS) messages. If the phone has a memory card, Cardtrp drops the Win32 PC virus known as Wukill onto the card.[8] Two proof-ofconcept Trojans, Crossover and Redbrowser, further show how widespread attacks could simultaneously hit desktops and mobile devices.[9] Both Trojans can infect certain mobile devices from PCs.

SMS, MMS, Bluetooth, and the synchronization between computers and mobile devices are all examples of potential attack vectors that extend the capabilities of malicious actors. Inherent vulnerabilities exist in modern mobile device operating systems that are similar to those of PCs and may provide additional exploitation opportunities. For example, the most recent Apple security update for iPhone OS 3.1.3 provided fixes for scenarios where playing a maliciously crafted mp4 audio file, viewing a maliciously crafted Tagged Image File Format (TIFF) image, or accessing a maliciously crafted File Transfer Protocol (FTP) server could result in arbitrary code execution.[10]

To help mitigate malicious activity affecting known vulnerabilities, users should install security patches and software updates as they become available.

## SOCIAL ENGINEERING

One of the more common methods of spreading malware on the Internet is through social engineering. Most malicious activity is often successful because users are deceived into believing it is legitimate. Exploitation by social engineering is extremely lucrative and will likely significantly increase in the mobile market.

Phishing is the criminal act of attempting to manipulate a victim into providing sensitive information by masquerading as a trustworthy entity. This technique is a well-established, significant cyber threat, and mobile devices provide unique opportunities for phishing, including variants such as vishing and smishing.

Vishing is the social engineering approach that leverages voice communication. This technique can be combined with other forms of social engineering that entice a victim to call a certain number and divulge sensitive information. Advanced vishing attacks can take place completely over voice communications by exploiting Voice over Internet Protocol (VoIP) solutions and broadcasting services.[11] VoIP easily allows caller identity (ID) to be spoofed, which can take advantage of the public's misplaced trust in the security of phone services, especially landline services. Landline communication cannot be intercepted without physical access to the line; however, this trait is not beneficial when communicating directly with a malicious actor.

Smishing is a form of social engineering that exploits SMS, or text, messages. Text messages can contain links to such things as webpages, email addresses or phone numbers that when clicked may automatically open a browser window or email message or dial a number. This integration of email, voice, text message, and web browser functionality increases the likelihood that users will fall victim to engineered malicious activity.

Regardless of the communication medium, users must ensure that any exchange of information occurs between their intended parties. Links contained in suspicious or unsolicited emails and text messages should be avoided, and to help prevent disclosing sensitive information to an unintended party via voice communication, users can initiate the phone call to a known, trusted number.

## EXPLOITATION OF SOCIAL NETWORKING

Social networking sites, such as Twitter and Facebook, have become mainstays of electronic information sharing. Information sharing often occurs with an unwarranted, inherent trust among users, as they blindly share and accept data from unauthenticated parties. Uniform Resource Locators (URLs) are constantly being exchanged within social networks as users share items of interest. Since a Twitter user is limited to 140 characters when posting an update, sharing a brief statement accompanied by a traditional URL may be impossible. The capability to significantly shorten a URL is provided by several different websites and is often integrated in social networking applications to happen automatically. Shortened URLs are invaluable in this case because they allow a URL with 137 characters to be shortened to 17 characters. For example:

http://brainstormtech.blogs.fortune.cnn.com/2010/02/12/help-wanted-obamas-twittererfilibusterers-need-not-apply/?source=cnn_bin&hpt=Sbin

becomes http://u.nu/72q95.

These services provide value, but they also make cyber criminals' goals much easier to achieve. Since the original URL is completely replaced, a user cannot know the destination of the shortened link without clicking on the link. Legitimate URLs are indistinguishable from those that are malicious, providing phishers with an effective cover. This tactic could lure a victim into unwittingly downloading malware or visiting a fraudulent site. It is highly likely that unsuspecting users would not think twice before clicking on the URLs.

Over the course of 2009, Facebook and Twitter experienced a 112% and 347% increase in mobile users, respectively.[12] This growing trend in mobile social networking provides an avenue for the exploitation of mobile devices.

## MOBILE BOTNETS

A botnet is a set of compromised computers, or bot clients, running malicious software that enables a "botherder" or "botmaster" to control these computers remotely. A botherder or botmaster can design a botnet to perform certain actions, such as information stealing or launching a denial of service, and issues commands to the bot clients from a command and control (C2) server. Since mobile networks are now well integrated with the Internet, botnets are beginning to migrate to mobile devices, as seen with Ikee.B.

Due to their ability to support rich content, MMS messages have a body field where Extensible Markup Language (XML) messages can be hidden.[13] Waledac, a web-based Internet botnet, uses XML messages to communicate. Unlike with Internet communication, Internet Protocol (IP) addresses are not used when exchanging SMS or MMS messages. Instead, mobile devices have an International Mobile Subscriber Identity (IMSI) and Mobile Subscriber Integrated Services Digital Network Number (MSISDN). These numbers are used to authenticate, register, and identify mobile network subscriptions by mapping the device to a phone number. The IMSI is embedded in the device hardware or contained on a removable card such as a Removable-User Identity Module (R-UIM) card in Code Division Multiple Access (CDMA) networks or a Subscriber Identity Module (SIM) card in Global System for Mobile Communications (GSM) networks. The MSISDN represents a phone number and is used to route communication to the subscriber. Domain Name System (DNS) also does not exist on mobile networks, making the use of advanced networking techniques such as fast flux and multi-homing impossible in mobile networks.[14] However, since mobile devices can have constant connections to the Internet, they can potentially be utilized like any other computer while maintaining all of their functionality within a mobile network.

Mobile devices using the Internet may be assigned dynamic private IP addresses that are inaccessible from the Internet, preventing a botmaster from communicating directly with a compromised host. Web-based botnets circumvent this obstacle by having bot clients poll web servers for further instructions. Any additional obstacles presented by using SMS or MMS messages to communicate could also be circumvented by adapting a web server to accommodate SMS and MMS functionality by creating a proxy that understands this type of communication and has a connection to the Internet. The capability to run a web server on the iPhone has existed since at least mid-2007.[15]

Compromised text messaging services could have severe consequences. In the aftermath of the recent earthquakes in Haiti, reputable charity organizations experienced a massive surge in text message donations. For example, a mobile device user could donate $10 to the American Red Cross by texting HAITI to 90999. In less than 48 hours, donations reached $5 million and accumulated at a rate of $200,000 per hour.[16] A mobile botnet could be configured to send text messages to a donation number set up for nefarious purposes. The donations could be small enough that a victim may not recognize the extra charge on his or her bill. The same concept could potentially be exploited in voting scenarios that leverage mobile devices or to carry out distributed denial of service attacks.

## EXPLOITATION OF MOBILE APPLICATIONS

Mobile applications, commonly called apps, provide enhanced convenience and functionality. Developers have created myriad mobile applications for various uses and activities, which is contributing to the proliferation of modern mobile devices. Anyone can potentially develop and distribute mobile applications with little oversight, making apps a potential attack vector for cyber criminals.

Several major banking institutions provide legitimate mobile applications that allow customers to conveniently check balances, pay bills, transfer funds, or locate automated teller machines (ATMs) and banking centers. However, banks are not the only ones creating banking-related apps. In early 2010, Google found potentially fraudulent banking applications in their Android Market. An anonymous developer known as "09Droid" sold a collection of banking applications that were not authorized by the banks for which they were seemingly developed.[17] It is unclear if the apps were used to gain access to users' confidential banking information. 09Droid published applications for approximately 40 different banking institutions, all of which Google removed from the Android Market.[18]

A similar incident occurred when Symbian unwittingly distributed the Sexy Space mobile worm as a legitimate, digitally signed application.[19] This malware steals subscriber, device, and network information from victims and has the capability to build a botnet. It propagates via spam text messages that are sent from a compromised device to the victim's contacts. The messages, exchanged at the expense of the victims, contain a link to a website hosting malicious applications that will infect the phone if executed. Currently, the Sexy Space mobile worm affects only Symbian mobile devices.

The validation and approval process for mobile applications varies by vendor. The following table provides a brief description of the policies of some of the more popular vendors.

| Vendor | Application Store | Application Development Policy |
|---|---|---|
| Apple | App Store | Apple requires developers to enroll in the iPhone Developer Program. Every application submitted to the App Store is evaluated by at least two reviewers for bugs, instabilities, unauthorized content, and other violations.[20] |
| Google | Android Marketplace | No requirements exist for publishing applications in the Android Marketplace. Once developers register, they have complete control over when and how they make their applications available to users.[21] |
| Microsoft | Windows Marketplace for Mobile | Developers must register with Windows Marketplace for Mobile. All applications sold on Windows Marketplace for Mobile must meet technical standards, be code signed, and pass policy checking and geographic market validation before they can be certified.[22] |
| RIM | Blackberry App World | Developers must create a vendor account to submit applications to the Blackberry App World. RIM reviews all submitted applications for content suitability and performs technical testing to ensure applications abide by the Blackberry App World Vendor Guidelines.[23] |
| Symbian | Horizon | Symbian Horizon is a publishing program and directory of Symbian Signed applications. To publish applications here, developers must obtain a Publisher ID and run the full Symbian Signed Test Criteria on applications before they can be made publicly available.[24] |

Many applications are regularly submitted to vendors for use on these platforms, including some that are malicious. Currently, the Apple App Store contains over 100,000 applications and receives about 10,000 new submissions each week. Apple has received applications that will steal personal data or are otherwise malicious and has rejected them during the review process.[25] As the volume of applications rises, it could be difficult to maintain high confidence in their integrity, regardless of the platform or policy.

## EXPLOITATION OF M-COMMERCE

M-commerce, or mobile e-commerce, is another growing trend with mobile devices. Consumers can use mobile devices from any location to research product information, compare prices, make purchases, and communicate with customer support. Retailers can use mobile devices for tasks such as price checks, inventory inquiries, and payment processing. For example, Apple Retail Store employees use modified versions of the iPod Touch that allow them to scan barcode labels and accept credit card payments from customers.[26]

The ability to read credit cards with a mobile device is not limited to retailers alone. A quick search for "credit card" in the Apple App Store reveals a number of different applications for accepting credit card payments. Third-party iPhone attachments for swiping credit cards are also available. "Square" is a small device that plugs into the iPhone's headphone jack and can transfer credit card swipe information to the supporting application. It also allows users to authorize payments in real-time via text message.[27] The Mophie "marketplace" is another credit card reader for the iPhone that will be available soon.[28]

Smartphones' credit card reader functionality has the potential to enable criminal activity such as "skimming" and "carding." Skimming is the theft of credit card information using card readers, or skimmers, to record and store victims' data. This activity is often accomplished in conjunction with otherwise legitimate transactions. Carding is the process of testing the validity of stolen credit card numbers. It can be done on websites that support real-time transaction processing to determine if the credit information can be successfully processed. The capability of a single compact hand-held device to perform each of these tasks will further enable malicious intentions.

## CONCLUSION

The user's limited awareness and subsequent unsafe behavior may be the most threatening vulnerabilities for mobile devices. It is critical to understand that a mobile device is no longer just a phone and cannot be treated as such. Unlike the previous generation of mobile phones that were at worst susceptible to local Bluetooth hijacking, modern Internet-tethered mobile devices are susceptible to being probed, identified, and surreptitiously exploited by hackers from anywhere on the Internet.[29] Many mitigation techniques for mobile devices are similar to those for PCs. US-CERT recommends the following best practices to help protect mobile devices:

- Maintain up-to-date software, including operating systems and applications;

- Install anti-virus software as it becomes available and maintain up-to-date signatures and engines;
- Enable the personal identification number (PIN) or password to access the mobile device, if available;
- Encrypt personal and sensitive data, when possible;
- Disable features not currently in use such as Bluetooth, infrared, or Wi-Fi;
- Set Bluetooth-enabled devices to non-discoverable to render them invisible to unauthenticated devices;
- Use caution when opening email and text message attachments and clicking links;
- Avoid opening files, clicking links, or calling numbers contained in unsolicited email or text messages;
- Avoid joining unknown Wi-Fi networks;
- Delete all information stored in a device prior to discarding it; and
- Maintain situational awareness of threats affecting mobile devices.

Anti-virus software exists for some mobile devices, which is one component of a layered defense. However, it can only assist in protecting against known threats. Users need to understand the threats and proactively take steps to avoid them. A high degree of vigilance is necessary to successfully prevent and mitigate future threats to mobile devices.

## ADDITIONAL RESOURCES

- US-CERT Cyber Security Tip ST06-007 – Defending Cell Phones and PDAs Against Attack
- US-CERT Cyber Security Tip ST05-017 – Cybersecurity for Electronic Devices
- US-CERT Cyber Security Tip ST04-020 – Protecting Portable Devices: Data Security
- US-CERT Cyber Security Tip ST06-001 – Understanding Hidden Threats: Rootkits and Botnets
- US-CERT – Virus Basics and Frequently Asked Questions

## REFERENCES

Apple Inc. About the security content of iPhone OS 3.1.3 and iPhone OS 3.1.3 for iPod Touch. 2010. Retrieved February 3, 2010 from http://support.apple.com/kb/HT4013.

Gary Allen. Exclusive look at Apple's new iPod touch-based EasyPay checkout. 2009. Retrieved Feburary 17, 2010 from http://www.appleinsider.com/articles/09/11/03/exclusive look at apples new ipod touc h_based_easypay_checkout.html.

Bill Brenner. Proof-of-concepts heighten mobile malware fears. 2006. Retrieved February 17, 2010 from http://searchexchange.techtarget.com/news/article/0,,sid43_gci1171168,00.html.

Jesus Diaz. iPhone Can Now Serve Web Pages, Run Python, Open Source Apps. 2007. Retrieved February 13, 2010 from http://gizmodo.com/282139/iphone-can-now-serveweb-pages-run-python-open-source-apps.

Ken Dunham, et al. Mobile Malware Attacks and Defense. 2009. Burlington, MA: Syngress Publishing, Inc.

Philip Elmer-DeWitt. 40 staffers 2 reviews 8,500 iPhone apps per week. 2009. Retrieved February 23, 2010 from http://brainstormtech.blogs.fortune.cnn.com/2009/08/21/40-staffers-2-reviews-8500-iphone-apps-per-week/.

Amy Feldman. Haiti Earthquake Provokes Wave of Text Donations. 2010. Retrieved February 13, 2010 from http://www.businessweek.com/investor/content/jan2010/pi20100114_236518.htm.

Flexispy Ltd. FlexiSpy Homepage. 2010. Retrieved February 17, 2010 from http://flexispy.com/.

F-Secure. Warning On Possible Android Mobile Trojans. 2010. Retrieved February 13, 2010 from http://www.f-secure.com/weblog/archives/00001852.html.

F-Secure. Worm:iPhoneOS/Ikee.B. 2009. Retrieved February 16, 2010 from http://www.f-secure.com/v-descs/worm_iphoneos_ikee_b.shtml.

Google Inc. Android Market Homepage. 2010. Retrieved February 23, 2010 from https://www.google.com/accounts/ServiceLogin?service=android developer.

GSM Association. Global Mobile Broadband Connections Increase Tenfold Over The Past Year. 2008. Retrieved February 4, 2010 from http://gsmworld.com/newsroom/press-releases/2008/870.htm.

Arik Hesseldahl. Apple's Schiller Defends iPhone App Approval Process. 2009. Retrieved February 13, 2010 from http://www.businessweek.com/technology/content/nov2009/tc20091120 354597.htm.

International Telecommunication Union. The World in 2009: ICT Facts and Figures. 2009. Retrieved February 3, 2010 from http://www.itu.int/ITU-D/ict/papers/2009/Europe RPM presentation.pdf.

John Leyden. Sign mobile malware prompts Symbian security review. 2009. Retrieved February 23, 2010 from http://www.theregister.co.uk/2009/07/23/sms_worm_analysis/.

Mike Melanson. Twitter Sees 347% Growth in Mobile Browser Access. 2010. Retrieved March 23, 2010 from http://www.readwriteweb.com/archives/twitter_sees_347_growth_in_mobile_browser_ac cess.php.

Microsoft Corporation. Windows phone Homepage. 2010. Retrieved February 23, 2010 from http://developer.windowsphone.com/ Help.aspx?id=0e4efb7b-2e57-4ff5-b381- 117281fc903b.

mStation Corporation. Mophie Homepage. 2010. Retrieved February 17, 2010 from http://www.mophie.com/product-p/1125_mp-ip3g-blk.htm.

Cyrus Peikari. Analyzing the Crossover Virus: The First PC to Windows Handheld Cross-infector. 2006. Retrieved February 17, 2010 from http://www.informit.com/articles/article.aspx?p=458169&seqNum=3.

Phil Porras, et al. An Analysis of the IKEE.B (DUH) iPhone Botnet. 2009. Retrieved February 3, 2010 from http://mtc.sri.com/iPhone/.

Dan Raywood. Google finds apparently fraudulent banking applications on its Andriod Marketplace. 2010. Retrieved February 1, 2010 from http://www.scmagazineuk.com/google-finds-apparently-fraudulent-banking-applicationson-its-android-marketplace/article/161047/.

Research in Motion Limited. Blackberry App World FAQ Homepage. 2010. Retrieved February 23, 2010 from http://na.blackberry.com/eng/developers/appworld/faq.jsp.

Anne Ruste Flo and Audun Josang. Consequences of Botnets Spreading to Mobile Devices. 2009. Retrieved February 2, 2010 from http://nordsec 2009.unik.no/papers/RFJ2009-NordSec.pdf.

Square, Inc. Square Homepage. 2010. Retrieved February 17, 2010 from https://squareup.com/.

Symbian Foundation. Symbian Horizon FAQ Homepage. 2010. Retrieved February 23, 2010 from http://horizon.symbian.org/index.php?option=com_content&view=article&id=52.

## End Notes

[1] International Telecommunication Union. The World in 2009: ICT Facts and Figures. 2009. Retrieved February 3, 2010 from http://www.itu.int/ITU-D/ict/papers/2009/Europe_RPM_presentation.pdf.

[2] GSM Association. Global Mobile Broadband Connections Increase Tenfold Over The Past Year. 2008. Retrieved February 4, 2010 from http://gsmworld.com/newsroom/press-releases/2008/870.htm.

[3] International Telecommunication Union.

[4] Ken Dunham, et al. Mobile Malware Attacks and Defense. 2009. Burlington, MA: Syngress Publishing, Inc.

[5] F-Secure. Worm:iPhoneOS/Ikee.B. 2009. Retrieved February 16, 2010 from http://www.f-secure.com/v-descs/worm_iphoneos_ikee_b.shtml.

[6] Phil Porras, et al. An Analysis of the IKEE.B (DUH) iPhone Botnet. 2009. Retrieved February 3, 2010 from http://mtc.sri.com/iPhone/.

[7] Flexispy Ltd. FlexiSpy Homepage. 2010. Retrieved February 17, 2010 from http://flexispy.com/.

[8] Cyrus Peikari. Analyzing the Crossover Virus: The First PC to Windows Handheld Cross-infector. 2006. Retrieved February 17, 2010 from http://www.informit.com/articles/article.aspx?p=458169&seqNum=3.

[9] Bill Brenner. Proof-of-concepts heighten mobile malware fears. 2006. Retrieved February 17, 2010 from http://searchexchange.techtarget.com/news/article/0,,sid43_gci1171168,00.html.

[10] Apple Inc. About the security content of iPhone OS 3.1.3 and iPhone OS 3.1.3 for iPod touch. 2010. Retrieved February 3, 2010 from http://support.apple.com/kb/HT4013.

[11] Ken Dunham.

[12] Mike Melanson. Twitter Sees 347% Growth in Mobile Browser Access. 2010. Retrieved March 23, 2010 from http://www.readwriteweb.com/archives/twitter_sees_347_growth_in_mobile_browser_access.php.

[13] Anne Ruste Flo and Audun Josang. Consequences of Botnets Spreading to Mobile Devices. 2009. Retrieved February 2, 2010 from http://nordsec2009.unik.no/papers/RFJ2009-NordSec.pdf.

[14] Anne Ruste Flo and Audun Josang.

[15] Jesus Diaz. iPhone Can Now Serve Web Pages, Run Python, Open Source Apps. 2007. Retrieved February 13, 2010 from http://gizmodo.com/282139/iphone-can-now-serve-web-pages-run-python-opensource-apps.

[16] Amy Feldman. Haiti Earthquake Provokes Wave of Text Donations. 2010. Retrieved February 13, 2010 from http://www.businessweek.com/investor/content/jan2010/pi20100114_236518.htm.

[17] Dan Raywood. Google finds apparently fraudulent banking applications on its Android Marketplace. 2010. Retrieved February 1, 2010 from http://www.scmagazineuk.com/google-finds-apparentlyfraudulent-banking-applications-on-its-android-marketplace/article/161047/.

[18] F-Secure. Warning On Possible Android Mobile Trojans. 2010. Retrieved February 13, 2010 from http://www.f-secure.com/weblog/archives/00001852.html.

[19] John Leyden. Sign mobile malware prompts Symbian security review. 2009. Retrieved February 23, 2010 from http://www.theregister.co.uk/2009/07/23/sms_worm_analysis/.

[20] Philip Elmer-DeWitt. 40 staffers 2 reviews 8,500 iPhone apps per week. 2009. Retrieved February 23, 2010 from http://brainstormtech.blogs.fortune.cnn.com/2009/08/21/40-staffers-2-reviews-8500-iphoneapps-per-week/.

[21] Google Inc. Android Market Homepage. 2010. Retrieved February 23, 2010 from https://www.google.com/accounts/ServiceLogin?service=androiddeveloper.

[22] Microsoft Corporation. Windows phone Homepage. 2010. Retrieved February 23, 2010 from http://developer.windowsphone.com/Help.aspx?id=0e4efb7b-2e57-4ff5-b381-117281fc903b.

[23] Research in Motion Limited. Blackberry App World FAQ Homepage. 2010. Retrieved February 23, 2010 from http://na.blackberry.com/eng/developers/appworld/faq.jsp.

[24] Symbian Foundation. Symbian Horizon FAQ Homepage. 2010. Retrieved February 23, 2010 from http://horizon.symbian.org/index.php?option=com content&view=article&id=52.

[25] Arik Hesseldahl. Apple's Schiller Defends iPhone App Approval Process. 2009. Retrieved February 13, 2010 from http://www.businessweek.com/technology/content/nov 2009/tc20091120_354597.htm.

[26] Gary Allen. Exclusive look at Apple's new iPod touch-based EasyPay checkout. 2009. Retrieved Feburary 17, 2010 from http://www.appleinsider.com/articles/09/11/03/exclusive_look_at_apples_new_ipod_touch_based_easypay checkout.html.

[27] Square, Inc. Square Homepage. 2010. Retrieved February 17, 2010 from https://squareup.com/.

[28] mStation Corporation. Mophie Homepage. 2010. Retrieved February 17, 2010 from http://www.mophie.com/product-p/1125 mp-ip3g-blk.htm.

[29] Phil Porras.

In: Mobile Device Security
Editor: Willliam R. O'Connor
ISBN: 978-1-62417-254-0
© 2013 Nova Science Publishers, Inc.

*Chapter 3*

# STOLEN AND LOST WIRELESS DEVICES[*]

## *Federal Communications Commission*

### BACKGROUND

The theft of wireless devices, particularly smartphones, is sharply on the rise across the country, according to many published reports. The high resale value of these high-tech phones has made them a prime target for robbers and the personal information contained on the device that could be used by identity thieves. Below are several steps that you can take to better protect yourself, your device, and the data it contains, along with instructions on what to do if your device is lost or stolen.

**How to Safeguard Yourself against Wireless Device Theft**

- Consider your surroundings and use your device discreetly at locations in which you feel unsafe.
- Never leave your device unattended in a public place. Don't leave it visible in an unattended car; lock it up in the glove compartment or trunk.
- Write down the device's make, model number, serial number and unique device identification number (either the International Mobile

---

[*] This factsheet was released by the Federal Communications Commission on April 10, 2012.

Equipment Identifier (IMEI) or the Mobile Equipment Identifier (MEID) number). The police may need this information if the device is stolen or lost.
- Review your warranty or service agreement to find out what will happen if your phone is stolen or lost. If the policy is not satisfactory, you may wish to consider buying device insurance.

## How to Protect the Data on Your Phone

- Establish a password to restrict access. Should your device be stolen or lost, this will help protect you from both unwanted usage charges and from theft and misuse of your personal data.
- Install and maintain anti-theft software. Apps are available that will:
  - Locate the device from any computer;
  - Lock the device to restrict access;
  - Wipe sensitive data from the device, including contacts, text messages, photos, emails, browser histories and user accounts such as Facebook and Twitter;
  - Make the device emit a loud sound ("scream") to help the police locate it.
- Make your lock screen display contact information, such as an e-mail address or alternative phone number, so that the phone may be returned to you if found. Avoid including sensitive information, such as your home address.
- Be careful about what information you store. Social networking and other apps may allow unwanted access to your personal information.

## What to Do if Your Wireless Device Is Stolen

- If you are not certain whether your device has been stolen or if you have simply misplaced it, attempt to locate the device by calling it or by using the anti-theft software's GPS locator. Even if you may have only lost the device, you should remotely lock it to be safe.
- If you have installed anti-theft software on your device, use it to lock the phone, wipe sensitive information, and/or activate the alarm.
- Immediately report the theft or loss to your carrier. You will be responsible for any charges incurred prior to when you report the

stolen or lost device. If you provide your carrier with the IMEI or MEID number, your carrier may be able to disable your device and block access to the information it carries. Request written confirmation from your carrier that you reported the device as missing and that the device was disabled.
- If the device was stolen, also immediately report the theft to the police, including the make and model, serial and IMEI or MEID number. Some carriers require proof that the device was stolen, and a police report would provide that documentation.
- If you are unable to lock your stolen or lost device, change all of your passwords for email, banking and social networking accounts that you have accessed using your device.

For More Information

For information about other telecommunications issues, visit the FCC's Consumer and Governmental Affairs Bureau website at www.fcc.gov/consumergovernmental-affairs-bureau, or contact the FCC's Consumer Center by calling 1- 888-CALL-FCC (1-888-225-5322) voice 1- 888-TELL-FCC (1-888-835-5322) TTY; faxing 1-866-418-0232; or writing to:

Federal Communications Commission
Consumer and Governmental Affairs Bureau
Consumer Inquiries and Complaints Division
445 12$^{th}$ Street, SW
Washington, DC 20554.

In: Mobile Device Security
Editor: Willliam R. O'Connor

ISBN: 978-1-62417-254-0
© 2013 Nova Science Publishers, Inc.

*Chapter 4*

# STATEMENT OF JASON WEINSTEIN, DEPUTY ASSISTANT ATTORNEY GENERAL, CRIMINAL DIVISION, U.S. DEPARTMENT OF JUSTICE. HEARING ON "PROTECTING MOBILE PRIVACY: YOUR SMARTPHONES, TABLETS, CELL PHONES AND YOUR PRIVACY"[*]

Good afternoon, Chairman Franken, Ranking Member Coburn, and Members of the Committee. Thank you for this opportunity to testify on behalf of the Department of Justice regarding privacy and mobile devices.

Over the last decade, we have witnessed an explosion of mobile computing technology. From laptops and cell phones to tablets and smart phones, Americans are using more mobile computing devices, more extensively, than ever before. We can bank, shop, conduct business, and socialize remotely with our friends and loved ones instantly, almost anywhere. These devices drive new waves of innovation, personal convenience, and professional resources. They also present increasingly tempting targets for identity thieves, cyberstalkers and other criminals.

Last month, one study concluded that 64% of American cell phone users were using smart phones.1 The speed and scale of that growth makes the topic

---

[*] This is an edited, reformatted and augmented version of testimony given on May 10, 2011 before the Senate Judiciary Committee, Subcommittee on Privacy, Technology and the Law.

of this hearing particularly timely. As mobile devices penetrate our daily lives, it is appropriate to evaluate the effect that these new devices have on our safety and privacy. We must also ensure that the law provides sufficient resources to investigators and prosecutors who investigate and prevent crimes against Americans who increasingly conduct their lives using this new medium. I thank the committee for giving me the opportunity to address these issues.

# PROSECUTING CYBERCRIMINALS AND IDENTITY THIEVES

One of the Department of Justice's core missions is protecting the privacy of Americans and prosecuting criminals who violate that privacy. Americans today face a wide range of threats to their privacy, including risks from using mobile devices. Foreign and domestic actors of all types, including cyber criminals, routinely and unlawfully access data that most people would regard as highly personal and private. Unlike the government – which must comply with the Constitution and laws of the United States and is accountable to Congress, courts, and ultimately the people – malicious cyber actors do not respect our laws or our privacy. The government has an obligation to prevent, disrupt, and deter such intrusions.

Every day, criminals hunt for our personal and financial data so that they can use it to commit fraud or sell it to other criminals. The technology revolution has facilitated these activities, making available a wide array of new methods that identity thieves can use to access and exploit the personal information of others. Skilled hackers have perpetrated large-scale data breaches that left hundreds of thousands—and in many cases, tens of millions—of individuals at risk of identity theft. Today's criminals can remotely access the computer systems of government agencies, universities, merchants, financial institutions, credit card companies, and data processors to steal large volumes of personal information—including personal financial information. As Americans accomplish more and more of their day-to-day tasks using smart phones and other mobile devices, criminals will increasingly target these platforms.

The most significant threats are continuing to evolve, and now increasingly include threats to corporate data. A report just released by McAfee and Science Applications International Corporation confirms this trend in cyber crime. According to this report, which was based on a survey of more than 1,000 senior IT decision makers in several countries, "high-end" cyber criminals have shifted from targeting credit cards and other personal data to the intellectual capital of large corporations. This includes extremely valuable trade secrets and product planning documents. These threats come both from outside hackers as well as insiders who gain access to critical information from within companies and government agencies. As entities make their key proprietary information available via mobile platforms, so that users can access it wherever and whenever it is most relevant, criminals and other actors will attack those devices as well.

The kinds of criminals we are up against are organized, international, and profit-driven. For example, in October 2009, nearly 100 people were charged in the U.S. and Egypt as part of an operation known as Phish Phry—one of the largest cyber fraud cases to date and the first joint cyber investigation between Egypt and the United States. Phish Phry was the latest action in what FBI Director Mueller described as a "cyber arms race" where law enforcement must coordinate and collaborate in order to keep up with its cyber adversaries. The defendants in Operation Phish Phry targeted U.S. banks and victimized hundreds of account holders by stealing their financial information and using it to transfer about $1.5 million to bogus accounts they controlled. More than 50 individuals in California, Nevada, and North Carolina and nearly 50 Egyptian citizens have been charged with crimes including computer fraud, conspiracy to commit bank fraud, money laundering, and aggravated identity theft. Led by the FBI and the United States Attorney's Office for the Central District of California, this investigation required close coordination with state and local law enforcement, the Secret Service, and our Egyptian counterparts. In late March, five more people were convicted of federal charges for their roles in this phishing operation, bringing the total number of convictions to date to 46.

One increasingly common form of online crime involves the surreptitious infection of a computer with code that makes it part of a "botnet" – a collection of compromised computers under the remote command and control of a criminal or foreign adversary. Criminals and other malicious actors can extensively monitor these computers, capturing every keystroke, mouse click, password, credit card number, and e-mail. Unfortunately, because many Americans are using such infected computers, they are suffering from an extensive, pervasive invasion of privacy at the hands of these actors.

Just last month, the Department announced the successful disruption of the Coreflood botnet, an international botnet made up of hundreds of thousands of computers that had been infected by malicious software (often referred to as "malware"). The Coreflood malware allowed criminals to remotely control the infected computers in order to steal private personal and financial information from unsuspecting computer users, including users on corporate computer networks. Through a combination of civil and criminal authorities, including a temporary restraining order, the FBI seized the servers that the criminals used to control the botnet and set up a substitute "command and control" server. The Coreflood malware was programmed to automatically contact the Coreflood command and control servers for instructions on a routine basis; after FBI intervention, those requests were instead routed to the FBI's substitute server. The FBI then replied to bot queries with an "exit" command that put the bots to sleep and stopped them from collecting further private data and causing more harm to hundreds of thousands of unsuspecting users of infected computers in the United States. As I'll discuss later in my testimony, the Department is concerned that as mobile devices become increasingly capable, they will be integrated into such botnets, or used to control them.

## THE DEPARTMENT'S ORGANIZATIONAL RESPONSE

The Department has organized itself to aggressively investigate and prosecute cyber crime wherever it occurs, including in the context of mobile devices and smart phones. Investigating and disrupting cyber crimes and cyber threats is a priority for the United States Attorney community, and the Attorney General's Advisory Committee has a subcommittee dedicated to cybercrime and intellectual property enforcement issues. A nationwide network of 230 Computer Hacking and Intellectual Property (CHIP) Assistant United States Attorneys in our USAOs focuses on these crimes, in coordination with the Criminal Division's Computer Crime and Intellectual Property Section (CCIPS). CCIPS provides core expertise on these issues, prosecutes cutting edge cases and provides litigation assistance to United States Attorneys' Offices. CCIPS also provides resources such as manuals, and trains prosecutors across the country, often in conjunction with Assistant United States Attorneys. Department prosecutors also work closely with our law enforcement partners.

In FY 2008 through FY 2010, United States Attorneys' Offices brought approximately 4,000 identity fraud cases. In addition, many of the large scale fraud cases prosecuted by the Fraud Section of the Department's Criminal Division also included identity fraud conduct.

The Office of International Affairs (OIA) enhances international cooperation efforts by expediting the sharing of critical electronic evidence with foreign law enforcement partners and by marshaling efforts to secure the extradition of international fugitives. The Office of Enforcement Operations guides investigative policy in numerous areas, including approvals for wiretaps and policy relating to use of tracking devices. It is a combination of these resources both in Main Justice and in the United States Attorneys' Offices that enables prosecutors across the country to tackle these complex and demanding cases.

The FBI Cyber Division is addressing the cybercrime threat from mobile devices through the Financial Threat Focus Cell (FTFC) and the Telecommunications Initiative. Through the FTFC the FBI Cyber Division is working with the largest U.S. based Financial Institutions (FIs) to determine the types, dates and level of mobile banking that those FIs are implementing. The FTFC is also working with FI organizations such as the FS-ISAC, BITS Financial Services Roundtable - Remote Channel Fraud Subgroup and the National Cyber-Forensics & Training Alliance's (NCFTA) Telecommunications Initiative. These organizations provide insight to the FBI so that law enforcement is more cognizant of current and future trends in terms of mobile banking product releases, new business alliances (e.g. AT&T, Verizon and Discover Card's recent product) and new mobile banking vendor companies.

In addition to the FI aspect to the mobile banking threat, the FTFC is working with the telecommunications sector through the Telecommunications Initiative (TI). As a part of the TI, the FBI is working with telecommunication organizations such as the Communication Fraud Control Association (CFCA) and the CTIA – The Wireless Association to address mobile banking and other telecommunications fraud matters. Through the relationship between both the FIs and the TIs the FBI has been able to develop fraud matters such as remote call forwarding, phishing fraud matters and telephonic denial of service (TDOS) attacks against high net worth FI customers. The FBI has ongoing relationships with a number of FI and TI partners to help organize the proactive sharing of fraud information to help mitigate or prevent economic loss. Furthermore, the FBI is beginning to share real-time intelligence with its international law enforcement (LE) partners in regards to global mobile threats. Finally, the FBI is proactively working with several anti-virus companies to stay on the forefront of mobile virus attacks and vulnerabilities.

The Department's work, and the work of our law enforcement partners, has helped to deter national and transnational cyber crime. The Verizon 2011 Data Breach Investigations Report, which is a joint study produced by Verizon, the U.S. Secret Service and the Dutch High Tech Crime Unit, found that more cyber crime investigations were conducted in 2010 than in any previous year, and concluded that the successful prosecution of identity thieves and other cybercriminals was having a significant impact. The report's leading hypothesis, in fact, was that "the successful identification, prosecution, and incarceration of the perpetrators of many of the largest breaches in recent history is having a positive effect."

## CYBER CRIME IN THE MOBILE CONTEXT

As mobile devices become more prevalent, identity thieves and other cybercriminals will begin to target the users of these devices. In fact, this may already be happening. In March, it was widely reported by technology researchers and journalists in the Washington Post, the New York Times, and elsewhere, that more than 50 apps for the Android mobile operating system had been modified to invade user privacy. According to the reports, these modified apps, infected by malware dubbed "Droid Dream," secretly installed malicious code on the device in addition to their apparent functions. This secret malware enabled the apps to steal sensitive information from the device, receive instructions from the criminals who had made the initial modifications, and even update their malicious capabilities. This activity is an example of the migration of criminal malware attacks that have targeted personal computers for years to targeting smart phones and mobile devices. As cell phones functionality expands, the line between mobile devices and personal computers becomes thinner. For criminals, this raises the prospect of millions of new sources of valuable personal and financial data, and millions of new devices to infect with malware and transform into "bots."

For acts that are particularly egregious – such as blatant theft of financial information or the malicious installation of malware I just described – criminal liability seems both appropriate and warranted. The Department of Justice has extensive experience with investigating and successfully prosecuting criminals who distribute malware and profit from their operation. It is the policy of the Department not to comment on ongoing criminal investigations, but criminal prosecution may be the most appropriate response to deter acts of this type and severity.

When deciding whether to bring an indictment under the Computer Fraud and Abuse Act (18 U.S.C. § 1030) ("CFAA"), Department prosecutors consider a wide range of factors, including the particular facts of the case, the law of the applicable circuit, the severity of the conduct, and the needs of justice. As mobile devices and services offered to mobile device users continue to expand, it will be important to distinguish between those cases that warrant criminal prosecution and those that may be best resolved through regulatory action. For certain less egregious actions, civil enforcement by the Federal Trade Commission might be more appropriate than criminal prosecution.

In addition to collection, it is also important to consider communications providers' ability to disclose the data that they collect from their customers. In this regard, it is important to note that under current law, communications providers may voluntarily disclose or sell any non-content data – such as information about a user's location – for any reason without restriction to anyone other than state, local, and federal government agencies. The Electronic Communications Privacy Act (ECPA) provides a broad exception for covered providers to disclose appropriately collected customer information to "any person other than a governmental entity." 18 U.S.C. § 2702(c)(6). This exception was included in ECPA at a time when there was great concern over ensuring the flexible development of the then-nascent Internet industry. As the commercial landscape changes, it will be important to ensure that our laws strike the appropriate balance and adequately protect consumers' privacy.

## CYBERSTALKING

One important consequence of the proliferation of mobile devices and services that collect location and other personal information about their users is the risk that stalkers, abusive spouses, and others intent on victimizing the user could use information from their mobile device to determine their whereabouts and activities. Stalking is not a new crime, and it is one that the Department of Justice takes very seriously. The increase in the use of mobile devices, however, raises new challenges that must be confronted.

The Department's Office on Violence Against Women (OVW) funds a number of projects that target the intersection of technology and the crimes of stalking, sexual assault, domestic violence, and dating violence. The Office recognizes that stalkers are increasingly misusing a variety of telephone, surveillance, and computer technologies to harass, terrify, intimidate, and monitor their victims, including former and current intimate partners. Perpetrators are also misusing technology to stalk before, during, and after perpetrating sexual violence. For young victims in particular, new technologies bring the risk of digital abuses such as unwanted and repeated texts, breaking into personal email accounts, and pressure for private pictures. Three OVW-funded projects, in particular, focus on "high-tech" stalking and the dangers that new technologies pose for victims.

First, for over ten years, OVW has funded the Stalking Resource Center, a program of the National Center for Victims of Crime, to provide training and technical assistance to OVW grantees and others on developing an effective response to the crime of stalking. The Stalking Resource Center has trained over 40,000 multi-disciplinary professionals nationwide, with an emphasis on the use of technology to stalk. Among other projects, the Resource Center has co-hosted nine national conferences that specifically focused on the use of technology in intimate partner stalking cases. In addition, with funding from the Department's Office for Victims of Crime, the Stalking Resource Center is currently developing two new training tools designed to help law enforcement officers, victim advocates, and allied professionals understand the most common forms of technology used by stalkers.

Second, since 2007, OVW has supported the National Network to End Domestic Violence's Safety Net Project, which works to identify best practices for using technology to assist victims. It is also concerned with training victim service providers to understand how stalkers may misuse technology and what strategies victims can use to increase their safety. In the past three years, the Safety Net Project has trained over 10,000 professionals and provided over 2,200 technical assistance consultations to OVW grantees and others.

Third, OVW funds the Family Violence Prevention Fund's "That's Not Cool" campaign to assist teens in understanding, recognizing and responding to teen dating violence. A critical part of this project is to help teens define their "digital line" as it relates to relationship and dating abuse. The website www.thatsnotcool.com was launched in January 2009 to help teens identify digital dating abuse and to encourage them to define for themselves what is and is not appropriate. So far the campaign has produced strong results, including over 900,000 website visits and 47,400 Facebook fans.

The Department has also strongly responded to the cyberstalking challenge through the prosecution of violations of the federal cyberstalking prohibition, 18 U.S.C. § 2261A. This statute allows for the prosecution of individuals who stalk using "the mail, any interactive computer service, or any facility of interstate or foreign commerce." This encompasses the use of the Internet through computers, smart phones and other mobile devices. Cases have been prosecuted under this statute based on conduct involving MySpace, Facebook and other social networking sites.

In one example of an egregious case charged under this statute, a defendant, posing as the victim, and using the victim's real name and address, posted photographs of the victim's children on a pornographic web site. Many men responded to this invitation.

The federal prohibition, however, is limited by the statutory requirement that the stalker and the victim be in different states, a requirement not found in other threatening statutes. This additional requirement may prevent prosecutors from charging cases, even where the conduct includes the most egregious acts. If an abusive spouse uses his spouse's phone to determine when she visits law enforcement for assistance, or to find where she is when she takes refuge with a friend, this may not violate 18 U.S.C. § 2261A as currently drafted because the two live in the same state. Similarly, a stalker from a victim's home town could potentially use location data from her phone to track her without violating the cyberstalking prohibition for the same reason. In fact, the case described in the previous paragraph was chargeable under 18 U.S.C. § 2261A only because the stalker and the victim, who met on the Internet, lived in different states. The Department is considering ways to address this limitation and looks forward to working with Congress on this issue. I hope that this Committee and Congress will take the necessary steps to ensure that law enforcement can continue to protect victims of cyberstalking, and deter their tormenters.

## INVESTIGATIVE RESOURCES FOR PROSECUTING COMPUTER CRIMES

Investigating and prosecuting multi-actor, multi-national crimes is extremely resource intensive. It is expensive to train and equip investigators and prosecutors to address the threat of cyber crime. As the proliferation of mobile devices provides criminals with new targets, the task of law enforcement will only get more demanding. Ensuring that law enforcement has the resources it needs to prosecute these crimes is a vital component to ensuring the safety and privacy of Americans.

For more specific details of the Department of Justice's needs for the coming year, I would direct you to the President's 2012 proposed budget, which outlines our detailed requests. In particular, the budget includes a request for funding for the Department to establish six Department of Justice Attaché positions that would emphasize the investigation and prosecution of laws prohibiting international computer hacking and protecting intellectual property rights at embassies around the world. Because computer crime is so often transnational in nature, it is vital that the Department have strong overseas representation to ensure that we can work more quickly and effectively with our international partners when investigating and prosecuting international computer crimes that target American citizens. The program would establish Department representatives at hotspots for computer and intellectual property crime around the world, and would help ensure that we can continue to protect American citizens' privacy, both at home and abroad. I hope that Congress will provide the resources that we need to establish this program and expand our resources to fight international computer crime.

## ENHANCING CRIMINAL INVESTIGATIONS AND PROSECUTIONS

In addition to the resource demands of combating cyber crime, law enforcement must have the authority to collect electronic evidence to investigate privacy invasions and protect public safety. One key statute that addresses this need, while also ensuring a fundamental balance between privacy and public safety, is the Electronic Communications Privacy Act. ECPA empowers law enforcement to collect the evidence it needs to prosecute a wide range of crimes. Department of Justice attorneys regularly use ECPA to obtain crucial evidence from mobile devices for all manner of investigations, including terrorism, drug trafficking, violent crime, kidnappings, computer hacking, sexual exploitation of children, organized crime, gangs, and white collar offenses. But it is important to understand that it plays a central role in the investigation of criminal invasions of privacy as well. When considering how best to protect the privacy of American citizens, I would ask that the Committee remember the important role that law enforcement plays in protecting Americans from privacy threats, and how ECPA is critical to our ability to continue to pursue that role.

One particular area of concern for the Department in collecting digital evidence – and one which bears directly on this hearing's topic – is ensuring that law enforcement can successfully track criminals who use their smart phones to aid the commission of crimes. When connecting to the Internet, smart phones, like computers, are assigned Internet Protocol (IP) addresses. When a criminal uses a computer to commit crimes, law enforcement may be able, through lawful legal process, to identify the computer or subscriber account based on its IP address. This information is essential to identifying offenders, locating fugitives, thwarting cyber intrusions, protecting children from sexual exploitation and neutralizing terrorist threats – but only if the data is still in existence by the time law enforcement gets there.

In my recent testimony in January before the House Judiciary Subcommittee on Crime, Terrorism, and Homeland Security, I outlined some of the serious challenges faced by law enforcement in this area in the more traditional computer context. ISPs may choose not to store IP records, may adopt a network architecture that frustrates their ability to track IP assignments and network transactions back to a specific account or device, or may store records for only a very short period of time. In many cases, these records are the only evidence that allows us to investigate and assign culpability for crimes committed on the Internet. In 2006, forty-nine Attorneys General wrote to Congress to express "grave concern" about "the problem of insufficient data retention policies by Internet Service Providers." They wrote that child exploitation investigations "often tragically dead-end at the door of Internet Service Providers (ISPs) that have deleted information critical to determining a suspect's name and physical location."

In one heart-wrenching example of the harm that a lack of data retention can cause, an undercover investigation that discovered a movie depicting the rape of a two-year-old child that was being traded on the internet was stymied because the ISP that had first transmitted the video had not retained information concerning the transmitter. Despite considerable effort, the child was not rescued and the criminals involved were not apprehended.

These challenges are equally serious in the context of smart phones and mobile devices. As the capabilities of smart phones expand, law enforcement increasingly encounters suspects who use their smart phones as they would a computer. For example, criminals use them to communicate with confederates and take other actions that would ordinarily provide pivotal evidence for criminal investigations. Just as some ISPs may not maintain IP address records, many wireless providers do not retain records that would enable law enforcement to identify a suspect's smart phone based on the IP addresses collected by websites that the suspect visited. When this information is not stored, it may be impossible for law enforcement to collect essential evidence.

In addition to collecting electronic evidence, it is vital to the success of the Department's mission that the scope and definition of criminal offenses is broad enough to allow us to prosecute the wide range of cybercrimes that are developing in today's increasingly networked society. This is particularly the case in the mobile context, where rapidly developing technology and services continue to provide opportunities for criminal acts. Some of the most egregious acts of privacy invasion that may be perpetrated on the users of mobile devices certainly rise to the level of criminal action under the CFAA. These include the installation of malware, theft of financial and personal information, and similarly severe acts, some examples of which I mentioned earlier. The Department takes these crimes very seriously, and, where criminal prosecution is warranted, is committed to vigorously prosecuting offenders. To date, we have not experienced shortcomings in the CFAA vis-à-vis mobile devices. We are continuing to review these authorities but do not have any particular proposals at this time.

\* \* \*

I appreciate the opportunity to share with you information about some of the challenges the Department sees on the horizon as Americans' use of smart phones and tablets continues to grow, and how the Department works to protect the privacy of users of mobile devices. We look forward to continuing to work with Congress as it considers these important issues.

This concludes my remarks. I would be pleased to answer your questions.

## End Note

[1] *March Mobile Mix Report*, Millennial Media, available at http://www.millennialmedia.com/research/mobilemix/.

## Chapter 5

# STATEMENT OF JUSTIN BROOKMAN, DIRECTOR, CONSUMER PRIVACY, CENTER FOR DEMOCRACY AND TECHNOLOGY. HEARING ON "PROTECTING MOBILE PRIVACY: YOUR SMARTPHONES, TABLETS, CELL PHONES AND YOUR PRIVACY"[*]

Chairman Franken, Ranking Member Coburn, and Members of the Subcommittee:

On behalf of the Center for Democracy & Technology (CDT), I thank you for the opportunity to testify today. We applaud the Chairman's leadership in examining the privacy issues presented by location-enabled mobile devices and appreciate the opportunity to address the lack of legal protection facing of what is one of the fastest growing areas of technological innovation.

CDT is a non-profit, public interest organization dedicated to preserving and promoting openness, innovation, and freedom on the decentralized Internet. I will briefly note the particular privacy issues presented by mobile services, and then describe the inadequacy of existing law to protect consumers. CDT strongly believes that legislation based on the full range of Fair Information Practice Principles (FIPPs) should be enacted to address the privacy challenges faced in the mobile space.

---

[*] This is an edited, reformatted and augmented version of testimony given on May 10, 2011 before the Senate Judiciary Committee, Subcommittee on Privacy, Technology and the Law.

## 1. THE PROMISE AND PERIL OF LOCATION-ENABLED MOBILE DEVICES

Mobile phones and tablets have exploded in popularity in recent years, and all evidence indicates that this trend will continue. Smartphone sales are expected to eclipse those of desktop and laptop computers combined in the next two years.[1] However, mobile devices store and transmit a particularly personal set of data. These devices typically allow third parties to access personal information such as contact lists, pictures, browsing history, and identifying information more readily than in traditional internet web browsing. The devices also use and transmit information consumer's precise geolocation information as consumers travel from place to place.

At the same time, consumers have less control over their information on mobile devices than through traditional web browsing. While third parties, like ad networks, usually must use "cookies" to track users on the web, they often get access to unique — and unchangeable — unique device identifiers in the mobile space. While cookies can be deleted by savvy users, device identifiers are permanent, meaning data shared about your device can always be correlated with that device. As is the case with most consumer data, information generated by mobile devices is for the most part not protected by current law and may be collected and shared without users' knowledge or consent.

Consumers interact with their mobile devices by running applications, or "apps" (i.e., programs designed to run on mobile devices). The mobile apps ecosystem is robust and offers an ever-increasing range of functionality from games, music, maps, instant messaging, email, metro schedules, and more. Mobile apps may be preinstalled on the device by the manufacturer or distributor, or users can download and install the programs themselves from their operating system's "apps store" (like iTunes or the Android Market), or a third-party store (like Amazon). App developers range from large, multinational corporations to individuals coding in their parents' basements. Generally speaking, we have seen a vibrant and creative app market develop for mobile devices. Unfortunately, it can be hard to know what information these apps have access to and with whom they are sharing it.

Recent studies of this flourishing apps data ecosystem have unearthed troubling findings. A recent survey indicated that of the top 340 free apps, only 19% contained a privacy policy *at all*.[2] Last December, the Wall Street Journal investigated the behavior of the 101 most popular mobile apps, finding that more than half transmitted the user's unique device ID to third parties without the user's consent.[3] Forty-seven apps transmitted the phone's location.[4] One popular music app, Pandora, sent users' age, gender, location and phone identifier to various ad networks.[5] In sum, a small phone can leak a big amount of data.

Once an app has access to a user's data, there are usually no rules governing its disclosure, and no controls available to consumers to regain control of it. For the most part, once data leaves the phone, it is effectively "in the wild." It may be retained long after the moment of collection, and often long after the original service has been provided. App developers, advertisers, ad networks and platforms, analytics companies, and any number of other downstream players can share, sell, or unpredictably use data far into the future. Even insurance companies are eying data mined from online services for new predictive models.[6] In short, today's mobile environment provides a gateway into an opaque and largely unregulated market for personal data.

Location data is of particular concern. In recent years, the accuracy of location data has improved while the expense of calculating and obtaining it has declined. As a result, location-based services are an integral part of users' experiences and an increasingly important market for U.S. companies. Consumers like the convenience and relevance of location based services. Location data can be used guide you to the closest coffee shop or help you navigate an unfamiliar neighborhood. Your location can be leveraged to connect you with coupons or deals in your immediate vicinity. And new, innovative, and useful services are introduced daily.

People generally carry their mobile devices wherever they go, making it possible for location data be collected everywhere, at any time, and potentially without prompting. Understandably, many find the use of location data without clear transparency and control troubling. Research shows that people value their location privacy and are less comfortable sharing their location with strangers than with acquaintances, and want granular control over their location information.[7] Indeed, location data is especially sensitive information that can be used to decipher revealing facts or put people at physical risk. Location information could disclose visits to sensitive destinations, like medical clinics, courts and political rallies. Access to location can also be used in stalking and domestic violence.[8] Finally, as an increasing number of minors carry location-capable cell phones and devices, location privacy may become a child safety matter as well.

There are also questions and concerns about the collection, usage, and storage of data by mobile platform providers such as Apple and Google.

Because in many instances, these companies are the ones actually calculating your location (based on comparing the WiFi access points in range of your device with known databases), they may receive extremely detailed information about consumer activity, considerably more so than traditional computer operating systems. Although these companies typically assert that data they receive from consumers is anonymized and used merely to build out their databases of access points, these limitations are self-imposed. Furthermore, these platforms may store detailed location and other customer information on the phone itself, which could then be accessed by government officials, potentially without a warrant, malicious hackers, or merely the person who finds your lost phone at Starbucks.[9]

Mobile devices and the services they enable provide consumers with great benefit. But it is imperative that Congress provide a clear policy framework to protect users' privacy and trust. CDT strongly supports privacy legislation that implements the full range of Fair Information Practice Principles (FIPPs) across all consumer data and provides enhanced protections for sensitive information, such as precise geolocation, including enhanced, affirmative opt-in consent. Unfortunately, today's legal protections fall far short.

## 2. Existing Legal Protections for Mobile Device Information are Outdated, Inapplicable, or Unclear

A number of laws aim to protect electronic communications, including location information. Unfortunately, technology has far outpaced these statutory protections in both the commercial and government contexts. An update is long overdue.

Following is a summary of relevant laws and an analysis of their application to today's location-enabled mobile devices.

### A. The Telecommunications Act of 1996 and Cable Communications Policy Act of 1984 (CPNI Rules)

Through the Telecommunications Act of 1996, with subsequent amendments, Congress has prohibited a telecommunications carrier from disclosing customer proprietary network information (CPNI), including "information that relates to the . . . location . . . [of] any customer of a telecommunications carrier . . . that is made available to the carrier by the customer solely by virtue of the carrier-customer relationship" — except in emergency contexts or "as required by law or with the approval of the customer."[10] In short, Congress issued a minimal standard that prohibited carriers from releasing location and other customer information on a solely discretionary basis.

Fifteen years ago, these privacy rules were a groundbreaking development. At the time, telecommunications carriers served as the primary gatekeepers for location information. Data about a cell phone user's location was calculated within a *carrier's* network using signals sent by the phone to the *carrier's* service antennas. These traditional protections have been left behind as we move from voice (traditionally the purview of telecommunications carriers) to data (which is often not the prevue of telecommunications carriers).

In light of modern location technology, there are at least two major shortcomings of the CPNI statute and resulting Federal Communications Commission (FCC) rules:

1) The CPNI rules simply do not apply to new types of location technologies, applications, and services. More specifically, the CPNI rules do not cover methodologies that are independent of telecommunications carriers covered by the law (e.g., WiFi database lookups, cell tower database lookups, or unassisted GPS locations). Thus, when an iPhone or Android user installs a location-based application, the location data transmitted by the resulting service is very likely completely unregulated under the CPNI rules.

2) Even, when a telecommunications carrier is involved in providing a location based service, it may not be covered by the CPNI rules because the FCC has removed wireless broadband service from Title II of the Communications Act (to which the CPNI rules apply) and deregulated it. When the Commission issued its Wireless Broadband Order,[11] Commissioner Copps explained the fractured effect of the Order on the protection of location information under the CPNI rules.[12]

Thus, modern mobile devices leverage location services that are largely invisible to the telecommunications provider and thus very likely outside the scope of the law. Although Congress and then the FCC did extend CPNI rules to cover IP-enabled ⁻interconnected⁻ VoIP services,[13] that protection still only extends to voice service regulated under Title II. At best, the application of CPNI rules to carrier-provided location-based data services is a murky question; at worst, the CPNI rules provide no protection whatsoever.

Practically speaking, this creates some striking confusion. A consumer using a mobile phone today can be protected by the CPNI rules one moment and unprotected the next. For example, a user might place a phone call using the traditional Commercial Mobile Radio Service (CMRS). In this case, they could feel secure that the CPNI rules required their carrier to protection their information. After the call, they use an Internet-based app or location service that uses location data rendered apart from the telecommunications carrier. Here, the user is likely unprotected.

## B. The Electronic Communications Privacy Act (ECPA)

The Electronic Communications Privacy Act was passed in 1986 primarily to address the issue of government access (about which, see below). However, it also contains important limitations on how companies may voluntarily share with other companies customer communications. Most notably, the law prohibits certain companies from sharing the content of customer communications or records without their consent.[14] In theory, this might prohibit mobile operating systems or applications from sharing consumer data without permission. Unfortunately, ECPA, while a very important and forward-looking statute at the time it was passed, was not written with the mobile apps ecosystem in mind. As applied to the current mobile environment, ECPA as a limitation on inter-business sharing of consumer data is, at best, vague and uneven.

When discussing the kinds of mobile applications and services at issue here today, it is not even clear which parties are currently covered by ECPA. ECPA's coverage of stored communications extends only to two categories of services — electronic communications services (ECSs) and remote computing services (RCSs). An ECS is a service that permits users to send or receive communications information (defined in part as ⁻signs, signals, writing, images, sounds, data, or intelligence of any nature¹)[15] to a third party or parties, like an email service or a private bulletin board such as a restricted Facebook wall. Some apps and location-based services are ECSs, some are not, and some fall into a grey area. For example, a service that allows users to share their location with a specific group of friends or associates is likely an ECS, with the ⎺data or intelligence¹ communicated to friends being the combination of the user's identity and her location data. However, an app that allows a user to share his location with a restaurant chain solely to allow it to return the location of the nearest restaurant is likely not an ECS, because it does not provide a way to communicate with third parties The statute ultimately requires highly fact-dependent analysis on the ECS question.

Remote computing services are, if anything, even more murky. An RCS includes any service that provides to the public computer storage or processing. The limited case law developed around this definition has not clarified its boundaries. Courts have held that websites enabling certain commercial transactions are not RCSs, but have suggested that remote processing of user-collected or -generated data is likely to be covered. Almost any app that collects user location or personal data and sends it to a remote server for further processing could, theoretically, fall under the ambit of this provision. However, it is important to note that mobile operating systems — the entities that often generate consumer location information in the first place — likely do not qualify as either ECSs or RCSs, and thus ECPA offers no protections at all as to those companies.

Of course, even if an app were to fall under the ECPA's ambit, there would still be open questions about whether customer data constituted the ⎺content¹ of a communication subject to protection. If a consumer affirmatively sent a location request to an app maker to ask for a nearby bar or restaurant, ECPA could arguably restrict the transfer of that information to third parties because the consumer's location was the content of a customer-initiated communication. If on the other hand, the app accessed the user's location in the background merely in order to send to a third party to serve relevant advertising, such request probably would not be governed. Such a reading of the statute would however lead to the perverse result that a consumer's information is afforded greater protections when she affirmatively shares sensitive data, as opposed to when her data is shared without her knowledge or consent.

Though the issue is not the focus of the present hearing, it is important to note that legislation to clarify the standards for government access to that information should also remain a Congressional priority. While the Communications Assistance for Law Enforcement Act (CALEA) indicates what the standard for law enforcement access to location information *is not,* no statute indicates what the standard for law enforcement access *is.* CALEA provides that a pen register or trap and trace order[16] cannot be used to obtain location information, but that statute is silent on what the standard should be.[17]

There is a federal statute on tracking devices, but it does not specify the standard that law enforcement must meet in order to place such a device.[18] Most importantly, the Electronic Communications Privacy Act (ECPA),[19] which sets up the sliding scale of authority for governmental access to information relating to communications (ranging from mere subpoena to warrant), does not specify what standard applies to location information.

This has resulted in a mish-mash of confused decisions while courts struggle to find and apply a legal standard. It has led to sometimes arbitrary distinctions based on whether location information is sought in real time or from storage, the degree of precision in the location information sought, the period(s) during which location information is sought, and the technology used to generate the location information. Some courts[20] have adopted a ⌐hybrid theory[1] advanced by the Department of Justice, holding that location information is accessible to government *in real time* if it meets the standard for *stored* transactional information in Section 2703(d) of the Stored Communications Act.[21] Other courts have required a higher level of proof – probable cause – for law enforcement access to this prospective location information.[22] As one federal magistrate judge recently testified in front of the House Judiciary Committee, there is no comprehensible standard for magistrate judges to apply when the government requests access to cell site location data – just an incoherent array of competing court decisions.[23]

As the first few circuit court decisions to address governmental requests for location information of all types have started to come down, it is becoming clear that the courts have constitutional concerns with these requests. In August, the D.C. Circuit held that putting a device in place to engage in extended GPS tracking without a warrant violates the Fourth Amendment.[24] In September, the Third Circuit held that magistrate judges faced with a request from the government for cell site location information have discretion under ECPA to insist upon a showing of probable cause, in part because of the potential sensitivity of the information.[25] Both the confusion in the lower courts and the consternation in the appeals courts demonstrate that Congressional attention to these statutes is sorely needed.

Congress enacted ECPA in 1986 to foster new communications technologies by giving users confidence that their privacy would be respected. ECPA helped further the growth of the Internet and proved monumentally important to the U.S. economy. Now, technology is again leaping ahead, but the law is not keeping up. CDT — through its Digital Due Process coalition — has convened technology and communications companies, privacy advocates and academics to create four principles for reforming ECPA for the next quarter-century. One of those principles is that location information should only be accessed through the use of a warrant[26] and we believe Congress should enact legislation that imposes a warrant requirement. Though the larger ECPA reform effort is and should remain independent of the issues being discussed here today, CDT believes setting easily-understood privacy-protective standards for government access to location data is a critical component of ensuring the privacy of American citizens and the success of American technology service providers.

## C. The Computer Fraud and Abuse Act (CFAA)

The Computer Fraud and Abuse Act (CFAA) is a criminal statute that prohibits intentional trespass into and theft from protected computer systems.[27] It criminalizes, in relevant part, one who ⁻intentionally accesses a computer without authorization or exceeds authorized access . . . information from any protected computer.[28] In short, it's a law to prosecute malicious hackers.

The CFAA is a law design to combat egregious computer crimes and cannot, and should not, be a primary tool in protecting consumers' mobile privacy from data sharing for marketing or related purposes. In the past, there have been failed attempts to stretch the CFAA to cover contractual terms of service.[29] CDT has warned that these attempts come with troubling encroachments on civil liberties and freedom of speech.[30] Criminal sanctions for certain computer crimes might well deter bad actors and provide appropriate tools in extreme circumstances. However, it is a blunt instrument not designed to address mobile privacy challenges arising from commercial activity.

The mobile market is nascent and innovating quickly. Many mobile app developers are individuals or small startup companies. They might be amateur programmers, working with various prefabricated pieces of code and advertising solutions. They may or may not have expertise in privacy or relevant law. Criminal sanctions, including jail time, would be heavy-handed and would likely chill the innovation we see today.

## D. Federal Trade Commission Act and State Attorneys General

Absent any affirmative legal requirements provided by sectoral specific privacy laws (such as those governing health or financial data), the default privacy rule for most consumer data is set by the FTC Act's prohibition on unfair and deceptive trade practices.[31] Under this authority, the FTC has established some general precedents about what constitutes a deceptive or unfair privacy practice online, such as recent settlements against companies who offered deceptive and ineffective opt-out solutions, and against Google for sharing personal data with other Google customers in violation of previous representations as part of the Buzz product. While these cases are important, they also demonstrate that the FTC is generally limited under current law to bringing enforcement actions against companies that make affirmative misstatements about their own privacy practices. In the absence of a baseline federal privacy law that gives the FTC the tools it needs and establishes it as the lead law enforcement agency for privacy matters, consumer protections in the location privacy space will continue to fall short. State Attorneys General also have consumer protection mandates that allow them to pursue service providers that engage in unfair or deceptive trade practices. To date, however, perhaps due to the inherent limitations in their authority, relatively little attention has been paid at the state level to consumer privacy concerns.

## 3. THE NEED FOR CONGRESSIONAL ACTION

Given that the default rule for most consumer data — including sensitive location data — is merely that companies cannot make affirmative misstatements about the use of that data, CDT strongly supports the enactment of a uniform set of baseline rules for personal information collected both online and offline. Modern data flows often involve the collection and use of data derived and combined from both online and offline sources, and the rights of consumers and obligations of companies with respect to consumer data should apply to both as well. The mobile device space implicates many different kinds of data in a complicated ecosystem. Cramming more notices onto small screens is alone insufficient. We need a data privacy law that incentivizes and requires companies to provide clear and conspicuous notice to consumers about the use of their information and provides for meaningful control of that information. Moreover, companies should collect only as much personal information as necessary, be clear about with whom they're sharing information, and expunge information after it is no longer needed.

The Fair Information Practices (FIPPs) should be the foundation of any comprehensive privacy framework. FIPPs have been embodied to varying degrees in the Privacy Act, Fair Credit Reporting Act, and other sectoral federal privacy laws that govern commercial uses of information online and offline. The most recent formulation of the FIPPs by the Department of Homeland Security offers a robust set of modernized principles that should serve as the foundation for any discussion of consumer privacy legislation.[32] Those principles are:

- Transparency
- Purpose Specification
- Use Limitation
- Data Minimization
- Data Accuracy
- Individual Participation
- Security
- Accountability

For particularly sensitive data, such as health information, financial information, information about religion or sexuality, and — most relevant here — precise geolocation data, a legislative framework should provide for enhanced application of the Fair Information Practice Principles, including for affirmative opt-in consent for the collection and/or transfer of such information. Consumers understandably have greater concerns about the use and storage of such information, and the law should err against presuming a consumer's assent to share such information with others. Furthermore, as noted above, the laws governing government access to consumer data should be modernized to require a warrant to access sensitive location information.

## CONCLUSION

CDT would like to thank the Subcommittee again for holding this important hearing. We believe that Congress has a critical role to play in ensuring the privacy of consumers in the growing market of mobile devices and services. CDT looks forward to working with the Members of the Subcommittee as they pursue these issues further.

## End Notes

[1] Cecilia Kang, Smartphone sales to pass computers in 2012: Morgan Stanley analyst Meeker, THE WASHINGTON POST, November 11, 2010, http://voices.washingtonpost.com/posttech/2010/11/smartphone_sales_to_pass_compu.html.

[2] Mark Hachman, Most Mobile Apps Lack Privacy Policies: Study, PC MAGAZINE, April 27, 2011, http://www.pcmag.com/article2/0,2817,2384363,00.asp.

[3] Scott Thurm and Yukari Iwatani Kane, Your Apps are Watching You, THE WALL STREET JOURNAL, December 17, 2010, http://online.wsj.com/article/SB10001424052748704694004576020083703574602.html

[4] Id.

[5] Id.

[6] Leslie Scism and Mark Maremont, Insurers Test Data Profiles to Identify Risky Clients, THE WALL STREET JOURNAL, November 19, 2010, http://online.wsj.com/article/SB10001424052748704648604575620750998072986.html.

[7] See, e.g., Janice Y. Tsai, Patrick Kelley, Paul Drielsma, Lorrie Cranor, Jason Hong, Norman Sadeh, Who's viewed you?: the impact of feedback in a mobile location-sharing application, Conference on Human Factors in Computing Systems: Proceedings of the 27th international conference on human factors in computing systems (2009), http://www.cs.cmu.edu/~sadeh/Publications/Privacy/CHI2009.pdf; Sunny Consolvo, Ian E. Smith, Tara Matthews, Anthony LaMarca, Jason Tabert, and Pauline Powledge, Location Disclosure to Social Relations: Why, When, & What People Want to Share, CHI '05: Proceedings of the SIGCHI conference on human factors in computing systems (2005), www.placelab.org/publications/pubs/chi05- locDisSocRel-proceedings.pdf.

[8] See, e.g., Rob Stafford, Tracing a Stalker, Dateline NBC, June 16, 2007, http://www.msnbc.msn.com/id/19253352/.

[9] See Alexis Madrigal, What Does Your Phone Know About You? More Than You Think, THE ATLANTIC, April 25, 2011, http://www.theatlantic.com/technology/archive/2011/04/what-does-your-phone-know-about-you-more-than-youthink/237786/.

[10] 47 U.S.C. § 222.

[11] Appropriate Regulatory Treatment for Broadband Access to the Internet Over Wireless Networks, Declaratory Ruling, WT Docket No. 07-53, FCC 07-30, 2 (rel. Mar. 23, 2007).

[12] Id at 2 (carriers offering Title I services ¯appear[] to be entirely free, under our present rules, to sell off aspects of the customer[s'] call or location information to the highest bidder.¹).

[13] See 47 C.F.R. § 64.2001, et seq.

[14] 18 U.S.C. §§ 2702(a).

[15] 18 U.S.C. §§ 2510(12).

[16] A pen register/trap and trace order permits law enforcement to obtain transactional, non-content information about wire and electronic communications in real time, including numbers dialed on a cellular telephone and telephone numbers of calls coming into a cell phone. See 18 U.S.C. §§ 3121-3127.

[17] 47 U.S.C. § 1002(a)(2).

[18] 18 U.S.C. § 3117.

[19] 18 U.S.C. §§ 2510 et seq.

[20] See, e.g., In re Application of U.S. for an Order for Disclosure of Telecommunications Records and Authorizing the Use of a Pen Register and Trap and Trace, 405 F. Supp. 2d 435 (S.D.N.Y. 2005).

[21] The SCA, part of the Electronic Communications Privacy Act, is codified at 18 U.S.C. §§ 2701 et seq.
[22] See, e.g., In re Application for Pen Register and Trap/Trace Device with Cell Site Location Authority, 396 F. Supp. 2d 747 (S.D.Tex. 2005).
[23] See Electronic Communications Privacy Act Reform and the Revolution in Location Based Technologies and Services Before the H. Comm. on Judiciary Subcomm. on the Constitution, Civil Rights, and Civil Liberties, 111th Cong. (June 24, 2010) (statement of Stephen Wm. Smith, United States Magistrate Judge).
[24] U.S. v. Maynard, 615 F.3d 544 (D.C. Cir. 2010).
[25] In the Matter of the Application of the United States of America for an Order Directing a Provider of Electronic Communication Service to Disclose Records to the Government, 620 F.3d 304 (3d Cir. 2010).
[26] For more information on the Digital Due Process coalition and its principles, see Digital Due Process at http://www.digitaldueprocess.org.
[27] 18 U.S.C. § 1030.
[28] 18 U.S.C. § 1030(a)(2)(C).
[29] See generally, US v. Drew, Electronic Frontier Foundation, available at https://www.eff.org/cases/united-states-vdrew (last visited May 6, 2011).
[30] Id.
[31] The FTC Act, 15 U.S.C. §§ 41 et seq.
[32] U.S. Department of Homeland Security, Privacy Policy Guidance Memorandum, The Fair Information Practice Principles: Framework for Privacy Policy at the Department of Homeland Security, December 2008, http://www.dhs.gov/xlibrary/assets/privacy/privacy_policyguide_2008-01.pdf.

# INDEX

## A

abuse, 44, 74
access, 2, 7, 9, 11, 13, 15, 16, 17, 18, 19, 20, 22, 23, 24, 25, 26, 27, 28, 30, 31, 36, 43, 44, 49, 50, 51, 54, 57, 60, 64, 65, 68, 69, 80, 81, 82, 84, 85, 86, 87, 89
adults, 6
advertisements, 27
age, 81
agencies, vii, 2, 5, 10, 11, 12, 16, 26, 28, 29, 33, 36, 39, 43, 68, 69, 73
alters, 17
American Red Cross, 53
apples, 58, 61
assault, 73
assessment, 35
assets, 43, 91
AT&T, 8, 40, 71
attacker, 17, 18, 21, 22, 23, 27, 43
attitudes, 40
Attorney General, v, 67, 70
audit, 5, 41
authentication, 12, 19, 20, 24, 25, 26, 34
authenticity, 25
authority(s), 11, 12, 70, 76, 78, 86, 88
awareness, 2, 3, 10, 11, 29, 33, 34, 35, 36, 37, 38, 39, 40, 56, 57

## B

bank fraud, 69
banking, 8, 25, 48, 54, 59, 60, 65, 71
banks, 54, 69
base, 8, 32
beneficial effect, 37
benefits, 5, 12
blogs, 52, 58, 61
Bluetooth, 6, 10, 18, 23, 27, 42, 43, 49, 50, 56, 57
bots, 70, 72
Broadband, 58, 60, 83, 90
browser, 17, 51, 59, 60, 64
browsing, 17, 80
businesses, 9

## C

campaigns, 10
case law, 85
cell phones, 67, 72, 81
challenges, 3, 29, 32, 44, 73, 77, 78, 79, 87
children, 34, 50, 75, 76, 77
China, 13
citizens, 69, 87
civil liberties, 87
clients, 52, 53
coding, 80
coffee, 81

collaboration, 34, 37
commerce, 49, 55, 74
commercial, 18, 49, 73, 82, 85, 87, 89
communication, 17, 19, 23, 42, 43, 44, 50, 51, 53, 85
communication technologies, 42
Communications Act, 43, 83, 86
Communications Act of 1934, 43
community(s), 11, 16, 70
competition, 12
compliance, 26, 33
computer, 4, 6, 10, 12, 16, 18, 26, 35, 41, 42, 43, 48, 53, 64, 68, 69, 70, 73, 74, 75, 76, 77, 82, 85, 87
computer fraud, 69
computer systems, 68, 87
computer use, 70
computing, 5, 7, 13, 43, 48, 67, 84, 85, 90
conference, 90
confidentiality, vii, 1, 4, 26
configuration, 2, 26, 28
Congress, 68, 75, 76, 77, 78, 82, 83, 86, 89
connectivity, 4, 10, 44
consensus, 43
consent, 80, 81, 82, 84, 85, 89
conspiracy, 69
Constitution, 68, 91
consumer protection, 88
consumer reliance, vii, 1, 4
consumers, 2, 3, 4, 5, 6, 7, 8, 9, 12, 14, 19, 20, 21, 22, 24, 25, 27, 29, 31, 33, 34, 35, 36, 42, 48, 73, 79, 80, 81, 82, 87, 88, 89
cooperation, 33, 70
coordination, 3, 10, 29, 69, 70
correlation, 36
credentials, 18
crimes, 68, 69, 70, 73, 75, 76, 77, 78, 87
criminal activity, 56
criminal acts, 78
criminal investigations, 72, 77
criminals, viii, 2, 13, 15, 16, 30, 44, 47, 52, 54, 67, 68, 69, 72, 75, 76, 77
critical infrastructure, 10, 11, 16, 43
cryptography, 30
customer data, 85
customer relations, 83
customers, 30, 31, 42, 54, 56, 71, 73, 88
cybersecurity, vii, 2, 3, 5, 10, 11, 13, 15, 29, 30, 32, 33, 34, 35, 36, 37, 38, 39, 41, 42
cyberspace, 11

# D

data transfer, 26
database, 30, 42, 83
decision makers, 68
defendants, 69
denial, 43, 53, 54, 71
denial of service attack, 54
Department of Commerce, 3, 10, 40
Department of Defense, 3, 12, 40
Department of Homeland Security, 3, 10, 39, 89, 91
Department of Justice, v, 16, 67, 68, 72, 73, 75, 76, 86
destruction, 43
detection, 30, 50
detection system, 30
DHS, 2, 3, 10, 11, 24, 25, 26, 27, 28, 29, 33, 34, 35, 36, 37, 38, 39, 41
digital evidence, 76
disclosure, 43, 81
domestic violence, 73, 81
donations, 53
draft, 10, 38, 39, 40, 45
drawing, 41
drug trafficking, 76

# E

earthquakes, 53
eavesdropping, 17, 27
e-commerce, 55
ecosystem, 80, 84, 88
ECPA, 73, 76, 84, 85, 86
education, 10, 35
educational materials, 33, 34, 35
Egypt, 69
electronic communications, 82, 84, 90

# Index

Electronic Communications Privacy Act, 73, 76, 84, 86, 91
electronic systems, 16
e-mail, 13, 17, 18, 20, 25, 26, 27, 38, 64, 69
emergency, 11, 83
employees, 28, 56
encryption, 2, 12, 24, 25, 30, 31, 34
enforcement, 12, 69, 70, 71, 73, 75, 76, 77, 85, 88
engineering, 43, 49, 51
environment, 11, 81, 84
equipment, 6, 15
espionage, 15, 16
Europe, 59, 60
evidence, 5, 41, 70, 76, 77, 80
execution, 50
executive branch, 11
expertise, 70, 87
exploitation, 48, 50, 52, 76, 77
exposure, 27
extradition, 71

## F

Facebook, 52, 64, 74, 84
FBI, 69, 70, 71
fears, 58, 60
federal advisory, 11
federal agency, 10
Federal Communications Commission, v, 3, 11, 37, 39, 63, 65, 83
federal government, 35, 73
Federal Trade Commission Act, 87
financial, 18, 25, 49, 68, 69, 72, 78, 87, 89
financial data, 68, 72, 87
financial institutions, 68
firewalls, 22
flaws, 18
Fourth Amendment, 86
fraud, 16, 30, 68, 69, 70, 71
freedom, 79, 87
funding, 34, 74, 75
funds, 16, 54, 73, 74

## G

gangs, 76
GAO, vii, 1, 2, 9, 15, 16, 18, 21, 23, 25, 26, 28, 38, 42, 43
geolocation, 80, 82, 89
google, 58, 59, 60, 61
governments, 16, 44
GPS, 48, 50, 64, 83, 86
growth, 49, 59, 60, 67, 86
guidance, 5, 12, 24, 25, 26, 27, 28, 34, 40, 41
guidelines, 10, 34

## H

hacking, 16, 75, 76
Haiti, 53, 58, 60
health, 87, 89
health information, 89
history, 17, 48, 72, 80
homeland security, 16
homes, 26
host, 53
hotspots, 76
House, 4, 77, 86
House of Representatives, 4
human, 90
Hungary, 49
hybrid, 86
hypertext, 3, 20
hypothesis, 72

## I

ID, 51, 55, 81
ideal, 48
identification, 3, 19, 57, 63, 72
identity, viii, 16, 23, 51, 63, 67, 68, 69, 70, 72, 85
image(s), 43, 50, 84
incarceration, 72
individuals, viii, 15, 17, 19, 28, 34, 36, 43, 47, 68, 69, 74, 80, 87

industry, 2, 3, 11, 18, 29, 30, 31, 32, 33, 37, 38, 40, 44, 73
infection, 14, 69
information sharing, 52
information technology, 12, 26
infrastructure, 3, 26, 28, 38
injury, 12
institutions, 16, 54
integration, 51
integrity, vii, 1, 4, 26, 55
intellectual capital, 69
intellectual property, 16, 30, 70, 76
intellectual property rights, 76
intelligence, 16, 71, 84
internal controls, 43
international communication, 11
international law, 71
interoperability, 11
intervention, 70
intrusions, 68, 77
invasion of privacy, 69
invasions, 76
IP address, 44, 53, 77
issues, 12, 25, 29, 30, 33, 36, 37, 40, 41, 53, 65, 68, 70, 78, 79, 87, 89

## J

Java, 7, 43
journalists, 72
Judiciary Committee, 67, 79, 86
jurisdiction, 12

## L

landscape, 73
laptop, 4, 18, 80
law enforcement, 69, 70, 71, 74, 75, 76, 77, 85, 86, 88, 90
laws, 68, 73, 75, 82, 87, 89
lead, 10, 11, 27, 34, 85, 88
leadership, 44, 79
leakage, 26
legal protection, 79, 82
legislation, 79, 82, 85, 87, 89
light, 83
litigation, 70
local area networks, 26
location information, 17, 81, 82, 83, 85, 86, 87, 89, 90
logging, 17

## M

malicious software, 2, 13, 14, 18, 52, 69
malware, 2, 13, 14, 16, 17, 18, 19, 20, 21, 23, 24, 25, 27, 34, 43, 49, 50, 51, 52, 54, 58, 59, 60, 69, 72, 78
man, 17, 23
management, 26, 28
mapping, 53
market share, 6, 7, 8, 40, 48
marketing, 87
marketplace, 56, 59, 60
mass, 48
materials, 33, 34
matter, 81
media, 11
medical, 81
membership, 44
memory, 25, 50
messages, 13, 18, 25, 27, 31, 50, 51, 53, 54, 57, 64
methodology, 5
Microsoft, 55, 59, 61
migration, 72
minors, 81
mission(s), 11, 43, 44, 68, 77
misuse, 20, 22, 64, 74
MMS, 50, 53
mobile communication, 16
mobile interactions, vii, 1, 4
mobile phone, viii, 8, 12, 27, 35, 47, 48, 49, 56, 84
mobile security reports, vii, 2
mobile telecommunication, 6, 11
models, 32, 81
modifications, 21, 22, 28, 72
modules, 26

money laundering, 69
morale, 16
motivation, 49
multinational corporations, 80
music, 80, 81
MySpace, 74

## N

National Cyber Security Alliance, 3, 31, 44
National Institute of Standards and Technology (NIST), 3, 10
national policy, 43
national security, 16
National Strategy, 11, 43
network operators, 26
networking, 9, 14, 52, 53, 64
nodes, 9
nonprofit organizations, 34

## O

obstacles, 53
offenders, 77, 78
Office of Management and Budget, 3, 12, 40
officials, 5, 31, 32, 33, 34, 35, 36, 38, 39, 40, 45, 82
OMB, 3, 12, 40, 43
omission, 12
openness, 79
operating system, 2, 6, 7, 12, 13, 19, 21, 22, 50, 56, 72, 80, 82, 84, 85
operations, 21, 43
opportunities, viii, 47, 50, 51, 78
organize, 71
outreach, 34
oversight, 54

## P

parents, 80
password, 2, 13, 19, 24, 31, 32, 34, 49, 57, 64, 69

permission, 17, 84
permit, 22
perpetrators, 72
personal computers, viii, 47, 72
personal information, viii, 23, 30, 34, 47, 63, 64, 68, 73, 78, 80, 88
photographs, vii, 1, 4, 75
physical environment, 8
platform, 7, 43, 49, 50, 55, 81
playing, 50
police, 64, 65
policy, 11, 26, 28, 43, 55, 64, 71, 72, 80, 82
population, 44
portability, 48
precedents, 88
President, 11, 75
prestige, 16
prevention, 26
principles, 28, 40, 86, 89, 91
private information, 19
private sector, 2, 6, 10, 29, 32, 34, 35, 38, 40
private sector officials, 40
probability, 44
professionals, 74
profit, 69, 72, 79
programming, 43
progress reports, 30
project, 74
proliferation, 54, 73, 75
property rights, 76
protection, 2, 11, 13, 19, 23, 25, 31, 32, 34, 83, 84, 85, 88
public awareness, 3, 29, 34
public education, 31
public interest, 79
public key infrastructure, 3, 26
public safety, 11, 42, 76
public sector, 31
public service, 10
publishing, 55

## R

race, 69

radio, 6, 8, 9, 42, 43
rape, 77
reading, 85
real time, 86, 90
recommendations, 2, 11, 30, 32, 33, 38, 39
reform, 87, 91
regulations, vii, 2, 5, 40
relevance, 81
religion, 89
reputation, 43
requirements, 10, 26, 55, 87
resale, viii, 63
researchers, 72
resolution, 6
resources, 11, 19, 36, 41, 67, 68, 70, 71, 75, 76
response, 30, 72, 74
restaurants, 8
revenue, 13
rights, 76, 88
risk(s), vii, 1, 2, 4, 5, 10, 19, 20, 21, 22, 23, 24, 26, 27, 28, 33, 36, 39, 40, 43, 68, 73, 81
risk assessment, 28
root, 22, 49
rules, 22, 25, 28, 81, 83, 84, 88, 90

## S

safety, 11, 42, 68, 74, 75, 76, 81
Samsung, 6, 40
sanctions, 87
school, 34
scope, 5, 29, 77, 83
scripts, 16
Secret Service, 69, 71
Secretary of Commerce, 37, 38, 39
Secretary of Homeland Security, 37
secure communication, 26
security, vii, 1, 2, 3, 4, 5, 10, 11, 12, 13, 16, 18, 19, 20, 21, 22, 23, 24, 25, 26, 27, 28, 29, 30, 31, 32, 33, 34, 35, 36, 37, 38, 39, 40, 41, 42, 43, 44, 45, 50, 51, 58, 59, 60
security practices, 11, 27, 28, 36, 37, 40, 44
security threats, vii, 1, 4, 24, 29, 39

Senate, 67, 79
sensitivity, 86
servers, 53, 70
service provider, 74, 87, 88
services, 6, 8, 12, 13, 16, 18, 44, 51, 52, 53, 73, 78, 79, 81, 82, 83, 84, 85, 89, 90
settlements, 88
sexual violence, 73
sexuality, 89
showing, 86
signals, 8, 83, 84
signs, 84
skimming, 56
SMS, 50, 51, 53
SNS, 42
social network, 48, 49, 52, 65, 74
society, viii, 47, 48, 77
software, 2, 6, 7, 12, 13, 14, 15, 17, 18, 21, 22, 24, 25, 26, 28, 30, 41, 42, 48, 49, 50, 51, 52, 56, 57, 64, 69
software code, 7, 13
SP, 43
spam, 16, 18, 21, 25, 31, 35, 54
specifications, 7, 26
speech, 87
Sprint, 8, 40
spyware, 16, 17, 21, 25, 49
state(s), 3, 36, 43, 69, 73, 75, 88, 91
statutes, vii, 2, 5, 40, 75, 86
storage, 5, 27, 81, 85, 86, 89
structure, 26
subpoena, 86
subscribers, 48
suppliers, 44
supply chain, 16
surveillance, 73
synchronization, 50

## T

tablet computers, vii, 1, 4, 6, 9, 14
target, viii, 27, 48, 50, 63, 68, 72, 73, 76
technical assistance, 74
technical comments, 38, 39
techniques, 13, 16, 17, 53, 56

# Index

technological advancement, 11
technological advances, 48
technology(s), 2, 5, 10, 14, 26, 42, 43, 48, 59, 61, 67, 68, 72, 73, 74, 78, 82, 83, 86, 90
teens, 74
telecommunications, vii, 2, 5, 7, 8, 11, 42, 65, 71, 82, 83, 84
Telecommunications Act, 43, 82
telephone, 6, 8, 13, 17, 73, 90
telephone numbers, 90
terrorism, 76
terrorist attack, 43
terrorists, 15
testing, 55, 56
text messaging, 30, 48, 53
theft, vii, viii, 3, 16, 23, 29, 30, 32, 34, 35, 56, 63, 64, 65, 68, 69, 72, 78, 87
threats, vii, viii, 1, 2, 4, 5, 11, 13, 14, 15, 24, 27, 28, 30, 31, 32, 36, 41, 42, 47, 48, 57, 68, 70, 71, 76, 77
time frame, 37, 38
Title I, 43, 83, 84, 90
Title II, 43, 83, 84
trade, 69, 88
training, 28, 74
transactions, 2, 4, 20, 24, 25, 26, 27, 44, 56, 77, 85
transmission, 5, 20, 44
transparency, 81
transport, 31

## U

U.S. economy, 16, 86
uniform, 88
United, v, 1, 6, 7, 8, 13, 18, 40, 44, 47, 68, 69, 70, 71, 91
United Kingdom, 13
United States, v, 1, 6, 7, 8, 18, 40, 44, 47, 68, 69, 70, 71, 91
universities, 68
USA, 8, 40

US-Computer Emergency Readiness Team, 3, 33, 40

## V

validation, 26, 54, 55
vector, 54
Verizon, 8, 40, 71
victims, 54, 56, 73, 74, 75
videos, vii, 1, 4, 14
violence, 73, 74
violent crime, 76
virtual private network, 3, 26
viruses, 17, 21, 25, 35
voting, 54
vulnerability, 18, 21

## W

Washington, 42, 43, 44, 65, 72
web, 14, 17, 22, 27, 42, 44, 51, 53, 60, 75, 80
web browser, 22, 51
web pages, 17
weblog, 58, 60
webpages, 51
websites, 7, 8, 16, 17, 18, 19, 24, 27, 34, 39, 42, 44, 52, 56, 77, 85
White House, 43
Wi-Fi, 57
wireless cellular network, 8
wireless devices, viii, 9, 26, 35, 63
wireless networks, 8, 28
wireless personal area networks, 8, 9, 43
wiretaps, 71
working groups, 12
worldwide, 5

## X

XML, 53